ON THE ECOLOGY OF
COENOBITA CLYPEATUS IN CURAÇAO

ON THE ECOLOGY OF
COENOBITA CLYPEATUS IN CURAÇAO

with reference to reproduction, water economy
and osmoregulation in terrestrial hermit crabs

ACADEMISCH PROEFSCHRIFT

TER VERKRIJGING VAN DE GRAAD VAN DOCTOR IN DE
WISKUNDE EN NATUURWETENSCHAPPEN AAN DE
UNIVERSITEIT VAN AMSTERDAM, OP GEZAG VAN DE
RECTOR MAGNIFICUS DR. A. DE FROE, HOOGLERAAR IN DE
FACULTEIT DER GENEESKUNDE, IN HET OPENBAAR TE
VERDEDIGEN IN DE AULA DER UNIVERSITEIT (TIJDELIJK
IN DE LUTHERSE KERK, INGANG SINGEL 411, HOEK SPUI)
OP WOENSDAG 7 NOVEMBER 1973, DES NAMIDDAGS
TE 13.30 UUR

door

PETER ARNOUD WILLEM JACOBUS DE WILDE
GEBOREN TE HEEMSKERK

Springer-Science+Business Media, B.V. 1973

Promotor: PROF. DR. A. PUNT
Co-referent: PROF. DR. J. H. STOCK

Dit proefschrift verschijnt tevens in Studies on the Fauna of Curaçao and
other Caribbean Islands Vol. 44.

ISBN 978-94-017-6702-6   ISBN 978-94-017-6768-2 (eBook)
DOI 10.1007/978-94-017-6768-2

STUDIES ON THE FAUNA OF CURAÇAO AND OTHER
CARIBBEAN ISLANDS: No. 144.

ON THE ECOLOGY OF COENOBITA CLYPEATUS IN CURAÇAO

WITH REFERENCE TO REPRODUCTION, WATER ECONOMY AND OSMOREGULATION IN TERRESTRIAL HERMIT CRABS

by

P. A. W. J. DE WILDE

(Carmabi, Curaçao / Dierfysiologisch Lab., Univ. v. Amsterdam)

CONTENTS

GENERAL INTRODUCTION

Among Crustacea terrestrial species are unusual, and only small numbers of species from various taxa have been more or less successful in occupying the land. In doing so they deviated from the normal evolutionary path from the primitive marine environment through brackish and fresh water to marshy and land areas; in this case some groups went straight from sea to land.

For several reasons connected with locomotion, gas exchange, water regulation, ion regulation and osmoregulation and the extreme variability of the external environment, a terrestrial way of life presents many problems. When a direct step is made from sea to land, particularly under hot and arid climatological conditions, such problems occur simultaneously.

All kinds of transitional stages may be distinguished. Among the Decapoda in the Netherlands Antilles only two brachyuran species and a hermit crab are genuine land crabs. For many days these animals can stay in dry habitats without any visible uptake of water. They are adapted to live almost completely on fresh water and are able to roam far inland. For the development of the larvae, however, they are still dependent on the sea.

Earlier papers on the physiology and ecology of terrestrial decapod Crustacea are scarce. Mainly by the work of GROSS, during the last decade, the subject finally received the attention it deserved on account of its promising possibilities. A symposium on 'Terrestrial adaptations in Crustacea' (1968) held in 1967 did not include research on hermit crabs.

From an evolutionary point of view, land hermit crabs are considered to be a relatively young group. Systematic and morphological characters provide evidence in favour of this view. As far as known no fossils have ever been found which could be identified with certainty as belonging to this group.

During the author's stay in Curaçao from 1965 to 1968 (DE WILDE, 1969) special attention was paid to problems concerning water economy and osmotic regulation in *Coenobita clypeatus* (Herbst). The Caribbean Marine Biological Institute (Carmabi) at Willemstad offered adequate laboratory facilities. Field experiments were mainly carried out on the easternmost part of the island (Oostpunt). The lack of details on both ecology and major aspects of the life history of the species, such as reproduction, justified additional research.

Though at first the field research was only a side-line, Chapter I is mainly devoted to its results. In this chapter full stress is laid upon the way those crabs manage to keep alive in their inhospitable habitats in Curaçao. Chapter II gives further information on the ecology of *Coenobita*. It appeared from visits to Bimini (Bahamas), the Miami area, Florida Keys, Jamaica, Aruba, Bonaire, Trinidad and Barbados, that *Coenobita clypeatus* may be found in widely diverging habitats. All the same a habitat as studied on Curaçao is considered to be fairly typical for the species.

In Chapters III to V – on Water management, Shell water and Osmoregulation respectively – adjustment problems are discussed more closely, mainly based on results obtained by relevant laboratory experiments.

REMARKS ON THE SPECIES

The small family Coenobitidae, which is mainly restricted to the tropics, comprises two genera only: *Birgus* with a single species (*B. latro* L., the coconut- or robber-crab), and *Coenobita* (the land hermit crabs) with some seven species.

This paper mainly deals with *Coenobita clypeatus* (Herbst), the only species in the western Atlantic area. (For a description of the species see CHASE & HOBBS, 1969). Though the material is considered to be uniform all over the area, a great many specimens were seen on Bimini which could be distinguished from the Curaçao animals especially by the conspicuous yellow colour of ambulatories and major cheliped, and by the golden yellow hairs around the male gonopore. Among the many thousands of animals examined in Curaçao there were some brick-red colour varieties and occasionally an albino.

According to PROVENZANO (1959) the species' distribution ranges from Bermuda, southern Florida and the Bahamas, through the West Indies to southern Brazil (CHASE & HOBBS, 1969, think that the Brazilian records need verification). It occurs in the coastal regions on to several miles inland, including considerable altitudes, e.g. up to the summit of the 887 m high Mountain of Saba. The largest numbers, however, are found in the more arid parts of its distributional area. As far as known the species seems to avoid places with a high rainfall and a luxuriant vegetation; it is rare or absent along the 'green coast' of the South-American continent.

In the Leeward Islands of the Netherlands Antilles *Coenobita clypeatus* – locally called "soldaatje" (little soldier) – is a very common and conspicuous animal. It is also common in the Windward Islands of the Netherlands Antilles (St. Martin, Saba and St. Eustatius), situated about 800 km to the North.

Part of the material on which this study has been based is preserved in the Zoological Museum of the State University, Utrecht.

ACKNOWLEDGEMENTS

The author wishes to express his gratitude to Prof. Dr. A. PUNT for his guidance, aid and advice in connection with all phases of the work. Sincere thanks are also due to Prof. Dr. J. H. STOCK for his valuable comments on the manuscript, to Dr. W. VERVOORT for his critical reading of the manuscript, to Dr. L. B. HOLTHUIS for solving some problems of nomenclature and to Dr. I. KRISTENSEN for directing my attention to the group of terrestrial decapods and for introducing me to the Carmabi on Curaçao.

To Dr. F. CREUTZBERG and the staff of the Institute I offer my thanks for the hospitality I received during my stay on the island. Prof. dr. A. J. PROVENZANO Jr. of the University of Miami I thank for his interest and valuable discussion, and Mr. R. F. MATHEWSON, resident director of the Lerner Marine Laboratory, for his kind aid and hospitality.

I owe a debt of gratitude to Mr. RUBEN ANGEL, who assisted me during the

6

nocturnal wanderings on Oostpunt, and to Mr. J. P. MAAL, the owner of the estate, who gave me permission to enter his properties.

Appreciations are due to the management and my colleagues of the Netherlands Institute for Sea Research for their patience in offering me the required time and facilities.

Miss S. M. VAN DER BAAN translated the manuscript into English, Mr. C. MARTIS turned the summary into Spanish, and Mr. H. HOBBELINCK made the drawings and the laboratory photographs.

In particular I am indebted to Dr. P. WAGENAAR HUMMELINCK, and to Drs. LOUISE J. VAN DER STEEN, both from the Zoological Laboraty of the University of Utrecht, for smoothing the long and difficult path of preparing the manuscript for the press, and for reading the proofs.

This study has been made possible by financial support from the Netherlands Foundation for the Advancement of Tropical Research (WOTRO). I acknowledge the valuable contacts with Dr. J. H. WESTERMANN. A stay in Bimini and some short visits to Florida and Jamaica were supported by a grant of the U.S. Naval Research (O.N.R. nr. 552–07).

FIELDWORK

I. MIGRATION AND REPRODUCTION

COASTAL AND INLAND ANIMALS

The populations of *Coenobita clypeatus* observed in Curaçao may be roughly subdivided into two different ecological types: coastal and inland.

Typical coastal animals occupy the habitats along the sea coast and, in places, along the banks of the so-called inner bays and lagoons. Food is rather scarce on the shorelines, and numerous other crustaceans such as the quicker and more active semi-terrestrial grapsoid crabs, are formidable food competitors. Populations of coastal animals are characterized by the large percentages of very small and small specimens, while the larger ones are often badly housed, occupying shells which are too small for their inhabitants or shells with large holes, cracks or worn interiors. Shells of quite unsuitable types, coming from various species such as *Murex, Melongena, Thais, Latirus, Bulla,* etc. are also common. Moreover lots of shells are often attacked by boring sponges or are covered by thick crusts of calcareous algae. Coastal animals always look more or less shabby.

Coastal animals depend mainly on sea water for their water supply; water stored in the shells was found in only a negligible number of cases. As will be explained more fully hereafter, coastal animals are more dependent on the sea.

Typical inland animals, on the other hand, give an impression of prosperity. Small animals are relatively fewer in numbers. The majority of the occupied shells came from the species *Livona pica*

(L.) (Pl. Ia, IVa). Shells of *Livona* (= *Cittarium*) are certainly the preferred ones in Curaçao, as was proved in a choice-experiment on a small scale.

The inland habitats are characterized by the occurrence of fresh-water drinking spots, shady trees and much better food conditions. Inland animals fully depend on fresh water for their needs. Shell water was usually found to be present to a certain amount.

Both types, with all sorts of transitional stages occurring in be-tween, become distinctly separated in the dry season after long periods of drought. In fact the existence of the two types of popu-lations is largely based on the possession of well-fitting and reliable shells. Very small specimens of *Coenobita* and also the larger ones with ill-fitting shells are supposed to be unable, or at least to have a poor chance, of ever reaching the favourable inland habitats. Lack of shell water in the dry season makes travelling increasingly diffi-cult. Breakage of shells will be fatal in most cases, as the chance of finding a new one in the interior is extremely small, and lethal desiccation of the unprotected animals will be only a matter of hours. In the inland region animals are worse off after moulting than those on the coast which have a variety of shells at their disposal.

It will be evident that only properly equipped animals are able to invade the favourable free inland habitats and to occupy them successfully. In particular a good shell will provide the possibility for land hermit crabs to live in such an advanced stage of terrestrial life. Inland animals originate from coastal animal populations but on the other hand an inland animal may also go back to the stage of a coastal animal.

For the Pacific *Coenobita clupeatus* auct. (= *C. brevimanus* Dana) HARMS (1932) also distinguished between two types of animals, which he calls "Formenkreisen." The terrestrial form inhabiting the dry coral foreshores, e.g. in the Moluccas up to an altitude of 80–100 m, may be distinguished from the coastal form – living in the zone next to the line of low water – by a number of modifications, due to the transition from an aquatic to a terrestrial way of life. According to HARMS even morphological differences may be observed, for instance in the covering membranes of the statocysts. However, the observations described in the present paper, con-cerning the related species *C. clypeatus* (Herbst) raise some doubt as to the system-atic value of the differences described by HARMS.

SIZE CLASSES

Old specimens of *Coenobita clypeatus*, living in shells of *Livona* or *Astraea* species may be as big as a man's fist, reaching gross weights up to 500 g, whereas small, but already sexually mature animals in *Tudora*, *Tectarius* or other small and light shells only reach gross weights of about two grammes.

Both, for field observations and laboratory experiments, it proved useful to subdivide the animals into classes, on account of the widely divergent weights and sizes. In practice the subdivision into 5 different size classes – strictly gross weight groups – was most satisfactory.

SIZE CLASS	ABBREVIATION	GROSS WEIGHT OF *Livona* – ANIMALS
very small	v.s.	< 10 g
small	s.	10–20 g
medium	m.	20–50 g
large	l.	50–100 g
very large	v.l.	> 100 g

Though this division is based on weights of land hermit crabs in shells of the gastropod *Livona pica* (L.) only, this does not complicate matters, since the majority of the bigger animals occupies shells of this species and nearly all experiments concern hermit crabs in *Livona*-shells. Anyway, individuals in shells of *Astraea tecta* (Lightfoot) and *Astraea caelata* (Gmelin), respectively the second and third choice of larger *Coenobita* in Curaçao, also fit satisfactorily into the above mentioned system.

With some experience it is well possible, as was proved by repeated experiments, to distinguish at sight between *Livona* and *Astraea*-animals with relatively few mistakes.

MIGRATION AND REPRODUCTION

Since the migration of land hermit crabs is known to occur all over the Caribbean, it is very remarkable that the phenomenon has received so little attention from scientists. Scattered over the whole area there are natives, in particular among elderly fishermen, who can tell stories about the clattering noise of huge numbers of soldier crabs, processions of several hundreds of meters long, at bright moonlight or, as others will have it, at pitch-dark nights. In the Netherlands Antilles a traditional story is told about invaders who took flight after they heard – literally by putting their ears to the ground – that many soldiers (soldier crabs) were on the march.

Because of the soft abdomens of the animals, which seem to be excellent bait for fishing, most Antillean fishermen show much interest in these creatures and know when they are absent, or at least difficult to find, during several months of the year in those places where they are usually found. From our own experience we know that the experimental animals were hard to obtain from about July till October.

There is no doubt that the terrestrial hermit crabs of the Family Coenobitidae, which have all become adapted to life on land and often spend the major part of their life far – up to several miles – from the coast, reproduce by means of larvae liberated into the sea. Based on animals reared in the laboratory, following two hatchings in 1960 on 22 July and 24 August, and two more hatchings in 1961, both on 10 August, PROVENZANO (1962) gives a description of the complete larval development of *Coenobita clypeatus*. From his experiments it is clear that the development of the larvae takes place in normal seawater.

VERRILL (1908) reports the occurrence of *Coenobita clypeatus* on sand hills in the Bermudas, far from the coast. PEARSE (1915) already mentioned a procession of about 500 individuals, moving over an area of about 200 m square, apparently migrating from the ocean through the desert of Santa Marta (Colombia), on August 29, 1913.

In a – rather imaginative – popular publication CARPENTER & LOGAN (1945) give an eyewitness account of two occasions on which they observed a "jamboree" of Coenobitas at Mona Island, halfway Puerto Rico and Hispaniola. They described how towards the end of August ten thousands of animals struggled along to one sandy beach.

None of the collected female specimens on Dominica in January, February, March and June by CHASE & HOBBS (1969) were ovigerous.

All the aforesaid communications show that the events dealing with the reproduction of *Coenobita clypeatus* are to be expected in a period ranging from about the middle of July till the end of August.

LABORATORY EXPERIMETS

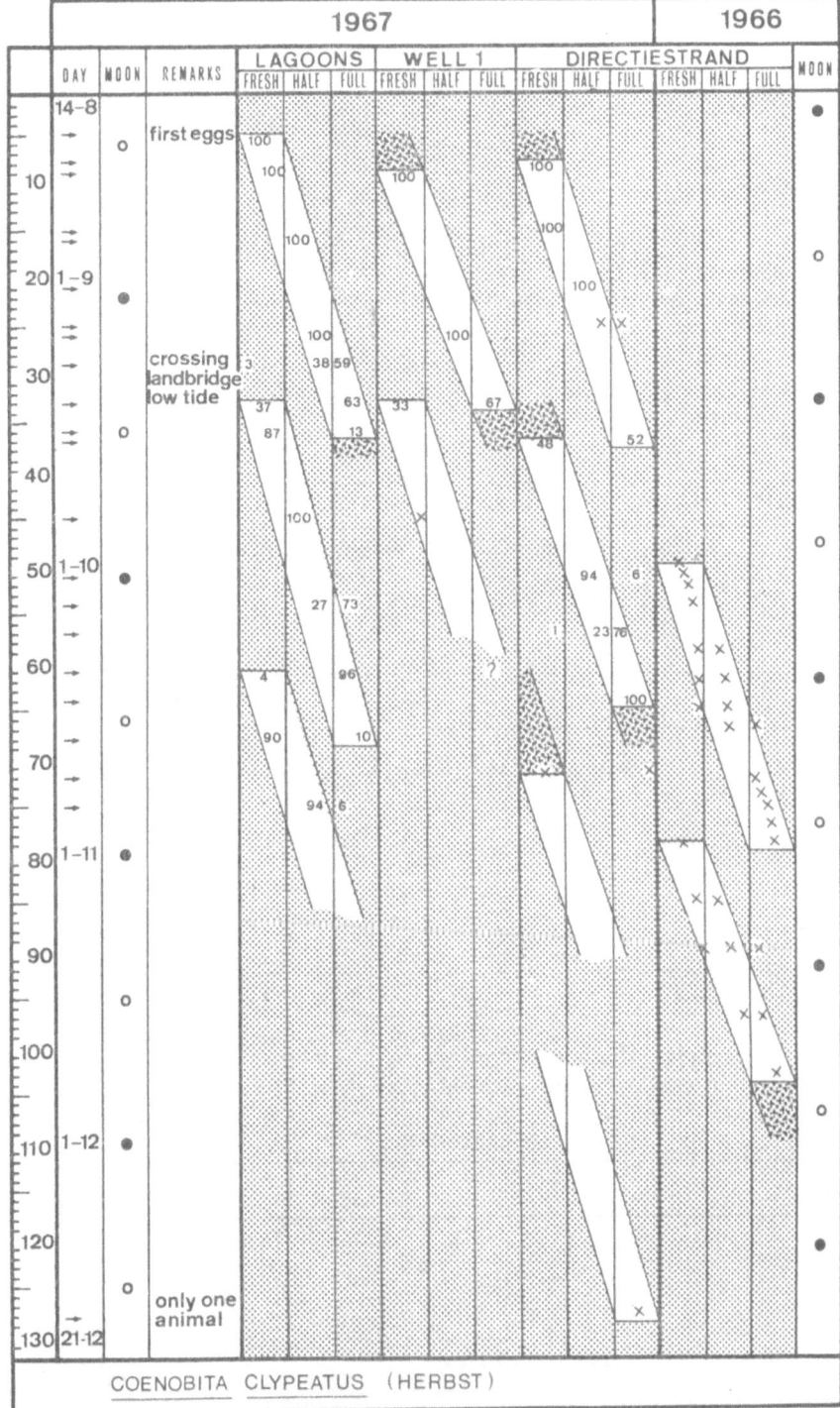

Fig. 5. SUCCESSIVE WAVES OF EGG PRODUCTION AND DEVELOPING EGGS at Oostpunt
(1967) and Directiestrand (1966/7). — Numbers: percentages of ovigerous females
with freshly spawned, half-developed and fully developed eggs; × × × : records
of which no percentages are known, or which are based on very small numbers;
dotted parts of bars: extrapolations.

TABLE 5

SEX RATIOS IN *Coenobita clypeatus* AS FOUND NEAR THE LAGOONS AT OOSTPUNT, 1967

Upper table: north of the Lagoons, at landside.
Lower table: south of the Landbridge, near the sea, at the time of egg-depositing, showing 100 per cent females.

Ratio = number of males divided by number of females, per size class.
Size classes: v.s. = very small, s. = small, m. = medium, l = large, v.l. = very large
Number of sexed individuals in italics.

a

	8.VIII	14.VIII	18.VIII	22.VIII	29.VIII	8.IX	11.IX	15.IX
v.s.	0.75 *7*	0.06 *50*	0.25 *50*	0.23 *53*	0.19 *50*	0.25 *25*	0.04 *25*	0.04 *75*
s.	0.25 *5*	0.14 *25*		0.56 *25*	0.39 *25*		0.75 *14*	0.67 *25*
m.	1.0 *14*			0.92 *25*	4.0 *25*		1.4 24	0.42 *17*
l.	3.0 *4*			7.3 *25*	6.5 15		5.5 13	
v.l.			11.0 *12*					

b

	11.IX	15.IX
v.s.		0.0 *50*
s.	0.0 *20*	0.0 *20*
m.	0.0 *21*	0.0 *14*
l.	0.0 *2*	0.0 *2*
v.l.		

a cont'd

	18.IX	27.IX	6.X	13.X	20.X	27.X	8.XII	29.XII
v.s.			0.14 *24*	0.27 *28*	0.23 *16*	0.19 *19*	0.75 *7*	
s.	1.22 *20*	0.47 *25*	0.38 *11*	0.0 *6*	0.13 *9*	0.0 *7*	0.37 *26*	0.0 6
m.	2.0 *30*	0.75 *28*	0.25 *25*	0.5 *18*	2.0 *12*	0.78 *16*	0.8 *9*	6.0 *7*
l.		∞ *4*	7.0 *8*	0.0 *2*		∞ *1*	3.0 *8*	
v.l.							∞ *1*	

b cont'd

	6.X	13.X	20.X	27.X
v.s.	0.0 *53*	0.0 *48*		
s.	0.0 *32*	0.0 *17*	0.0 *1*	0.0 *3*
m.	0.0 *16*	0.0 *22*		0.0 *1*
l.	0.0 *1*	0.0 *1*		
v.l.				

Field observations in the nights of August 22 and 29 and of September 8, and the results of subsequent sampling hardly produced any new data. There was still a clear increase of the number of animals present in the vicinity of the Lagoons in general, and in the surroundings of the Landbridge and the Delta area, in particular. An aggregation in these parts was also demonstrated by the movements of the orange (new) series, which were released on the eastside of Lagoon A. A fair number of these animals were recovered later on near Lagoon C. Estimations showed that the total number of animals near the Lagoons was now some ten thousand, perhaps even over one hundred thousand.

By now the colour of the eggs carried by a fair number of the females had changed gradually from dark reddish brown, through to light reddish brown, light grey and greyish blue. Dark eyespots indicated very distinctly the advanced development of the eggs.

Again all possible suitable parts of the coastline, both on the north and south coasts of Oostpunt were searched for other concentrations of hermit crabs. Apart from the concentrations already known to exist around Well 1, not exceeding 200 animals, nothing could be detected.

Just as on October 2, 1966, when masses of hermits on the Directiestrand had been noticed, weather-conditions on September 3, 1967 were equally dull, with a drizzling rain. Visiting the Directiestrand in the morning again thousands of very small animals, mostly living in *Tectarius* shells, were seen on the coral debris, limestone boulders and under the manchineel trees. Samples produced lots of ovigerous females with developing eggs, but nothing special could be detected.

On September 11 under quiet weather conditions (partly clouded, moon in its first quarter, low tide) thousands of hermit crabs, not previously observed in this spot, were seen struggling down over the Landbridge in the direction of the sea coast. Clusters of animals in which all sizes were represented were distributed between the tide marks over a distance of about 100 m. Here at low tide with little wave action a number of quiet, seawater-filled rockpools occur. Hundreds of hermits frequented the edges of the rockpools, under and against stones covered with algae and also on the still moist coral debris, but never, neither completely nor partly, submerged in the water. Clusters of fully developed eggs were attached to the stones above water level. Samples of water from the rock-pools and off the coast were found to contain young larvae.

No males were present among egg-depositing females in the rockpools, neither were there any crossing the Landbridge. Samples of animals from the Landbridge and the adjacent part of the coast showed sex ratios of 0.0 in all size classes as is shown in Table 5.

This phenomenon, namely that the majority of the ovigerous females with fully developed eggs separate themselves from the main group about the time of egg deposition, thus leading to distinctly lower sex ratios than when the animals are collected closer to the sea, was used later on to find evidence for other possible hatching places. Samples were taken on the eastside of Lagoon A and near Well 1, where hatching places were expected to exist. Sex ratios from these places, however, did not give any information to this effect.

In 1967 the phenomenon occurred again on the Directiestrand, though not as clearly as near the Landbridge.

Based on the observations on the Directiestrand in 1966 and in 1967 and also at Oostpunt near the Lagoons, near Well 1 and again on the Directiestrand the appearance of another group of females with freshly spawned eggs was expected and did indeed occur towards the end of the first wave of egg production and development, so that there must be a second wave (vide Fig. 5). Further observations showed the occurrence of a third, and at one time even of a fourth wave. The strongly decreased numbers of ovigerous females of which this third (and fourth) wave(s) were composed, made it impossible to follow the complete waves in detail.

Further visits to Oostpunt on September 15 and 18 showed a gradual decrease of the percentages of females with fully developed eggs and subsequently an increase of the percentages of females carrying freshly spawned eggs. On September 27 only fresh or partly developed eggs were found.

In the meantime, and following the end of the first wave, the total number of animals near the Lagoons decreased markedly (Fig. 4).

The second time that great activity on the Landbridge was noticed, was on October 6.

The same low sex ratios, showing almost 100 per cent females, were found in samples from the sea coast. Later on females with freshly spawned eggs were observed and a further decrease of the total of hermits near the Lagoons. During the second half of October only a few animals were left.

Observations of diffuse migrations in several places in the grounds of Oostpunt indicated the re-population of the inland. Though none of the recovered tags in the surroundings of Well 4 had been recorded earlier on the southcoast, the growing population near Well 4 was obvious enough (Fig. 4).

In addition, a fair number of well-sized animals was found in a small machineel grove, N.E. of Awa Blancu and about 1500 m west of the Landbridge. No animals had been observed here in the preceding months. Two recoveries of orange (new) tags on October 27 and December 29 in this place proved that the animals came from the Lagoon area. So did a yellow (new) tag, recovered on October 20 in a place midway between Wells 2 and 3 on the north coast.

During the last visits to the Lagoons within the scope of these experiments in December 1967 only a limited number of individuals could be counted. At that time, however, the population of Well 4 was obviously not yet at its full strength.

Visiting Oostpunt on June 25 1968 some fellow-workers of the Institute observed two tagged animals, a black and a red one, near Well 4. They did not observe whether they were originally tagged animals or whether moulting or changing of shells had occurred in the meantime. A similar observation near Well 4 followed in October 1968 at which occasion three tagged animals were seen; apart from a violet and a black tag also an orange (new) one from the lagoon was found.

Finally it must be recorded that in June 1970 an assistant of the Carmabi collected some 100 animals near Well 4. Three years after the original tagging experiments a quarter of the animals present was found to be still in possession of tagged shells.

DISCUSSION

From the tagging experiments described above and the chrono-logical field observations carried out in the second half of 1967 and the supplementary data from 1966 a satisfactory picture could be assembled concerning the seasonal migrations and other activities connected with reproduction of the land hermit crab *Coenobita clypeatus* Herbst in Curaçao. Though important details as copu-lation, sexual behaviour, etc., were not observed, the story as a whole has become sufficiently clear.

There are only few publications on migration and sexual behaviour of *Coenobita clypeatus* and closely related species and therefore the data described here cannot be compared with the findings of the other authors. For the present it must be kept in mind that some of the data recorded here may perhaps apply to the islands of the Leeward Group only and other comparable islands and coastal areas with a dry climate and some hinterland within the dispersal area of *Coenobita*.

Typical "coastal animals", at least as far as their reproduction is concerned, do not show any migrations as described here. Flourish-ing populations such as occur on the very small Keys off Kingston Harbour, Jamaica, e.g. Drunkenmans Cay which were visited in November 1967, have no possibillities for migration whatever. *Coenobita* living under quite different ecological circumstances, such as occur in Bimini (Bahamas) or Saba (Windward Group) may show patterns different from those discussed in this paper.

Reproductive season and mating places

Reproduction of *Coenobita clypeatus* in the Caribbean takes place in summer and autumn, dispersed over several months. In 1967 first indications of migrations in Curaçao were noticed early in July; the last ovigerous female was found on December 19.

Judging from the freshly spawned eggs in 1966 and 1967, first noticed on October 2 and August 18 respectively, considerable changes in time may occur.

Since PROVENZANO (1962) mentioned hatchings on July 22 and August 10 and 24, there must have been freshly spawned eggs at least since the end of June in the Florida region. It is possible that in the northernmost part of the dispersal area of this species, reproduction may even start some weeks earlier. The absence of any sign of eggs or larvae, during my visit to Bimini, early in November 1967, also points in that direction.

As *Coenobita* often lives in habitats far from the sea coast and its larvae need the open ocean for their development, the aim of the summer migrations is clear. They appear to migrate to preferential spots along the coast, which are not necessarily close to the shore. The striking feature is the directedness of the movement.

PEARSE (1915) mentions a migration of *Coenobita* consisting of about 500 individuals, moving over an area of about 200 square meters, away from the ocean in North Colombia. Eleven animals were collected and preserved; all animals proved to be females and 8 of them were carrying eggs. CARPENTER & LOGAN (1945) described mass movements of land hermit crabs to one sandy beach, measuring three acres, in Mona Island. In Saba (personal communications of Mr. A. J. GEERLINK, 1965) fishermen annually observe large numbers of hermit crabs descending certain gullies from the mountain slope. The present paper describes the migration of the large majority of the hermit populations living in the eastern part of Curaçao towards a small area on the south coast of the same island.

No doubt difficulties such as those caused by the steep cliffs, in combination with the extreme exposure to wave action along the north coast of Curaçao restrict suitable places for egg deposition by the females and further development of the young larvae; but it cannot be explained how the adult animals are aware of these complications. On the other hand sheltered inlets as occur e.g. in the St. Joris Baai on the north coast and in most parts of the south coast, with moderate wave action, including open inlets as Awa di Oostpunt and Fuik Baai seem to offer more suitable spots than the 100 m wide strip near the Landbridge.

Some preliminary experiments already showed a higher mortality of newly hatched larvae in hypersaline water: for example a combination of 125% sea water and a temperature of 28°C showed an almost 100% mortality within 24 hours. PROVENZANO (1962) mentions successful rearings of larvae under laboratory conditions in filtered sea water of a salinity of about 36.0‰ and a temperature of 29°C. Therefore, water of higher salinity in inlets and lagoons seems to be less suitable for the development of the larvae. There is no evidence that the larvae of *Coenobita* are equally well adapted to estuarine environments with steep salinity gradients as those of the brachyuran land crab *Cardisoma guanhumi* Latreille (KALBER & COSTLOW, 1968).

Since *Coenobita* was observed to regulate the salinity of its shell water by diluting it with fresh water if necessary, the presence of drinking places along the northern banks of the Lagoons seems to be the main factor why *Coenobita* prefers these particular localities. From the moment of arrival up to the time of liberation of the larvae by the females – a period of at least five weeks – which makes high demands upon the water management of the ovigerous females in particular, the animals have to remain in an arid and barren area. As far as known, the surroundings of the Lagoons are the only places along on Oostpunt to offer both quiet, clear sea water of the right salinity, and relatively fresh water, within easy reach for Coenobitas.

The large number of animals round about the drinking places show the need for fresh water, which unquestionably is a seepage of rain water originating from the limestone terraces above Ceru Port'i Camber and Ceru Palibandera or even perhaps from the Ceru Blancu table land. By mixing with the salt lagoon water, in the outcrop on the Lagoon banks just above the water level, it obtains its brackish character (see Table 4). Low salinities such as 1.6‰ and 1.8‰ from the samples taken on October 27 show that the available water is derived from fresh water. High water levels in

the Lagoons as observed on November 6 cause temporary inunda-
tions of some of the drinking places, which make them unsuitable
for *Coenobita*.

Except for Well 1 no other spawning places where ovigerous females could be observed were found on the north coast of Oostpunt,
though it has been searched thoroughly. Just as in the Wells 2 and 3
on the north coast suitable drinking water is offered here at a short
range from the ocean, however, the ocean is not by any means a
quiet water.

The development of the eggs near Well 1 went on smoothly.
Deposition of eggs on the nearby coast during the favourable nights
of September 15 and 18 could not be observed while at the same
time it was in progress on the South coast.

As stated before the population on the Directiestrand is intermediate between real "Coastal" and "Inland"-animals. The situation on the Directiestrand is hardly comparable with that at Oostpunt. Although fresh water is not permanently available, the water
conditions seem to be sufficient for reproducing *Coenobita*. Shady
trees and thick layers of mould and old leaves provide shelter.
Manchineel fruits and waste water apparently provide a sufficient
supply of moisture.

Migrations

From field observations and tagging experiments it must be
concluded that the migrations down to the Lagoons from Well 4 are
straight and directional. Not a single tag was recovered near the
seven other check points, all situated around Well 4; not a sign of
movements elsewhere than near Cer'i Palibandera and east of Awa
Blancu was noticed.

At present a random swarming, followed by a concentration by
trial and error in a suitable spawning place seems very unlikely.

Starting from a larval development of at least some weeks, of
which the greater part is planctonic, and also considering the strong
currents both in the hatching place and in this part of the Caribbean, we may suppose that part of the larvae will be passively

transported over considerable distances. Probably part of the animals in Curaçao will have come from the Bonairean coast or from more eastward islands (Aves Islands).

PROVENZANO (pers. comm.) believes, however, that at least part of the littoral larvae are retained in their hatching area and that the question of transport is not as simple as is generally supposed. For instance EMERY (1972) proved that on Barbados a re-stock-ment of the decapod population of shallow water takes place from locally produced larvae, in spite of strong and steady currents impinging upon that island from the East. Their taking off is pre-vented by a series of eddies and counter-currents, which previously were left out of consideration.

The colonization, more or less by chance, of the inland habitats by the larger and properly equipped animals, favoured by proper weather conditions, is quite understandable. On the other hand a direct migration to a fixed breeding place by a crustacean, hatched far away, transported passively by sea currents, moved upon the shore in any possible place and finding an appropriate habitat by chance only, requires a lot of our imaginative faculties on the point of homing and directional orientation of land hermit crabs. Obser-vations made on Oostpunt point to a guidance of young and un-experienced animals by experienced animals.

Whatever the mechanism of orientation may be, by some internal or external signal, migration towards the spawning places starts in the early summer in Curaçao and adjacent areas.

On July 5, 1967, when the first series of tagged animals were released near Well 4, migrations had probably already started. From the number of recovered black animals and the number of caught untagged animals on the next day, a very rough calculation may be made, giving some idea about the strength of the population present around Well 4. Assuming that the population did not change very much in the course of one day, and also that tagged and untagged animals occurred at random, the number was estimated at 7–800 individuals (very small animals not counted). A similar calculation from the recovered black animals on July 7 gave about twice as many of them present near Well 4. Another calculation based on recovered red animals on July 14, two days after the release, gave a number of over 400.

Recoveries of violet and yellow tagged animals near Well 4, released 500 m westward and eastward of the Well, proved that medium-sized animals moved over at least 500 m in a single night. Previous measurements, based on walking velocity over relatively short distances of about 50 m, gave theoretical values of about 1000 m per night. Assuming that the lower value is the correct one, hermits from Well 4 can cover the distance down to the Landbridge, the shortest distance being about 4000 m, in eight days, but probably in a shorter time. In theory they can do that without any water supply on the way. As a matter of fact such a journey will take several weeks at leisure, time spent in looking for food and water included.

A yellow animal released on July 5 was recovered on July 21 near Palibandera, thus covering 3 km in 16 days. The first observed tagged animal near the Lagoons, between Landbridge and Delta, also a yellow one, made a jouney of 40 days and the first of the red animals, released on July 12, which was observed near the Lagoons had taken 36 days to cover a distance of 4 km.

It is supposed that favourable weather and moist terrain conditions will highly stimulate migration.

As to the preferred route between Well 4 and the Lagoon-area we cannot say anything as yet. It seems plausible that the animals follow the escarpment of the Ceru Blancu in the direction of Cer'i Palibandera, and from there through the gully southwards; others perhaps will take Rooi Sjon Tata and go from Ceru Patía down to the coast. Hardly any activity was noticed on the track between Well 4 and Awa Blancu and the migrating animals did not show any interest in the fresh water basin of Poos Grandi along the track.

Only animals in good condition and fitted with a reliable shell migrate towards the spawning places. The great majority of the non-migrating hermits, observed in the inland habitats e.g. those of Well 4, proved to be of the smaller size classes or were obviously badly housed. Quite a few of the examined females proved to be deficient, which means that they had underdeveloped pleiopods.

Typical small-sized and shabby coastal animals as usually observed round Well 3 and in several other places, scattered along the coast line, not showing migration, were also found to have high percentages of deficient females as is shown in the two examples following:

1) A sample of 42 small- and medium-sized animals from the coast east of Fuikbaai, taken on August 29, proved to consist of 52.4% ♂♂, 14.3% normal ♀♀, and 33.3% deficient ♀♀.

2) Another sample of 60 small- and medium-sized animals from Well 3, which was taken on September 8, consisted of 56.7% ♂♂, 6.6% normal ♀♀, and 36.7% deficient ♀♀.

Tagging experiments with black, violet, yellow (old) and red series, as mentioned before, gave certainty about the migrations from Well 4 towards the Lagoons. Considering the vastness and inaccessibility of the area, the number of recoveries was not disappointing. In Table 6 the number of animals recovered in 1967

TABLE 6

RECOVERED TAGS OF COENOBITAS RELEASED NEAR WELL 4,
OOSTPUNT, 1967

In addition to the recovered totals, the data for recoveries near the Lagoons are
given separately.

series	number released	recovered total		recovered near Lagoons	
		number	percentage	number	percentage
black	168	46	27.4	8	4.8
yellow (old)	118	13	11.0	8	6.8
violet	130	12	9.2	4	3.1
red	184	35	19.0	7	3.8
total	600	106	17.7	27	4.5

divided into the total numbers and those from the Lagoons only, are summarized. Tags observed twice or more often are not included. It cannot be concluded from these figures that any of the colours should be more easily noticed in the field than the other colours; apparently the observation of tagged animals is mainly based on the recognition of a contrasting disk on the shell of the individuals.

Comparing the percentages of the recovered tags of all the series near the Lagoons it appears that they are within the same order of magnitude, and that they do not show any disadvantage for the violet and yellow (old) animals, which were released at a distance of 500 m from Well 4.

When studying the numbers of recovered animals, obvious differences appear between the various size classes.

Though perhaps for the relatively small numbers the conclusions are somewhat speculative, the figures of Table 7 seem to indicate reduced ability of the small animals to perform the trip to the spawning places on the coast. On the other hand, this would be also in entire conformity with the assumption of special inland populations, being characterized by the large percentages of animals in the larger size classes.

TABLE 7

RECOVERED COENOBITAS FROM THE LAGOONS DIVIDED INTO
SIZE CLASSES

Black, yellow, violet and red tags. Number of released
individuals in parentheses.

class	number	percentage
small	1 (105)	1.0
medium	11 (283)	3.9
large & very large	15 (212)	7.1
total	27 (600)	4.5

Mating

Animals in the Lagoons settle on the northern banks in the close vicinity of the drinking places, and in the course of time concentrate in the area between the Landbridge and Delta. Huge numbers of well-housed animals, possibly something like a hundred thousand, at the height of the season, lead us to assume the presence of numerous inland populations existing in the grounds of Oostpunt; the population of Well 4 being just one of them.

From this crowd of animals 27 different specimens out of a number of 600 tagged animals from Well 4, could be recovered. The number of tags observed more than once was low. Only once the same tag was recovered in a different place on a different day. Three other repeated observations were always made in exactly the same spot, under the same stone or in the same hole.

From the data obtained it is possible to make a rough calculation which percentage of the population of Well 4 succeeded in ac-complishing the tour to the coast. The small numbers of green and orange (old) tags found near the Lagoons indicate that in any case the chance to do this from Well 2 or 3 is much smaller.

At daytime the Lagoon banks are mostly empty, the animals hiding in the labyrinth of crevices and holes in the limestone deposits, or between the twisted roots of the numerous *Conocarpus* trees. At night they are active, refilling the shell-water and assembling in certain, more or less flat places on the bank, often together with animals of about the same size. Though copulation was not observed, it seems very likely that this will happen on these special "playing grounds." It is suggested that mating takes place between animals of roughly the same size group only. It seems impossible for sexually mature, very small animals, for the greater part females, living in *Tectarius*- or *Tudora*-shells, with a gross weight of 1 g to mate with giant animals up to a gross weight of 500 g. The rough division on the "playing grounds" into size groups which have no sharp limits and partially overlap, depends on the ability to succeed in mating. In particular groups of very small animals in *Tectarius*- and *Tudora*-shells, and in shells of the smaller *Nerita*-species are obvious. In other places groups of large and very large specimens are numerous, whereas congregations of medium-sized animals are most common. The demonstrated differences in sex ratio between smaller and larger animals indicate that there exists only a restricted supply of females in the larger size groups. This will certainly result in interesting differences in sexual behaviour within the various size groups.

Animals on the "playing grounds" often stay in the same place for a long time; they are active, have raised bodies, and continuously move the antennae and antennules. However, no attention was paid to special patterns of behaviour, associated with mating as describd by HAZLETT (1966a) for a number of marine hermit crabs of Curaçao. In his observations on the social behaviour of *Coenobita* (1966) no sexual behaviour is mentioned.

Small numbers of animals from the "playing grounds" observed in glass aquaria in the laboratory did not show any sign of actual mating.

The curious descriptions of CARPENTER & LOGAN (1945) may concern mating on Mona Island: "At sunset they scattered on the beach. Then came the startling event. The jamboree began and it was as though the grandfather of all hermit crabs had called through his megaphone: 'change shells with the lady on your right'. The hermit crabs came out of their portable houses and began to exchange them. It was moving day, dance marathon and bargain basement rolled into one. A good many were soon looking the worse for wear; but the brawl went on ... Less than a week after the start, the sun rose on an empty beach. The hermit crabs had all gone back into seclusion on the top of their fortress rock."

It is difficult to fit CARPENTERS observations into the sequence of events as they were observed on Curaçao.

On some occasions shell-fighting as described by HAZLETT (1966) was observed, but never the exchange of houses on a large scale. The rapid retreat of the hermits to the inland area of Mona Island is completely different from our observations.

In 1966 the activity on the "playing grounds" in Oostpunt lasted for several weeks, starting in the middle of August and terminating towards the middle of October. The peak of activity was observed some days before full moon. According to the activity measurements in the Laboratory (Chapter III) diurnal activity on the "playing grounds" and near the drinking places was also greatest in the early evening, slowing down towards midnight.

When very small and small females from the "playing grounds" are removed from the shell the swollen, dark reddish-brown gonads can often be seen through the transparent skin of the abdomen. In older females this is less obvious because of the thicker tegument.

Sometimes, by pressing the abdomen of a male, long spermatophores may be produced from the gonopores (Pl. Va); the latter are widely opened and the tips are covered with black or dark brown hairs. These sperma threads are whitish in colour and sometimes have thickenings like a rope of beads. They are very elastic and sticky, and in old males reach a length of over 10 cm.

After preliminary precopulatory behaviour the copulation itself, on the analogy of HAZLETTS observations in marine species, is supposed to be a partial moving out of the shell – perhaps as observed by CARPENTER. When gonopores of both animals are opposed, parts of the spermatophores will be transferred to the female, possibly with the aid of the fifth pereiopodes. Soon afterwards both animals will pull back, after which postcopulatory behaviour may be shown.

Egg production and egg development

The freshly spawned eggs are firmly attached to the thin but extremely tough hairs of the branched pleopodes at the left side of the abdomen (Pl. Vb).

With a freshly caught female, placed into an artifical house of transparent glass, the role of the modified fifth pereiopodes in attaching the eggs could clearly be seen (Pl. VIb).

In 1967 the first freshly spawned eggs were observed near the Lagoons on August 18; one day before full moon. Only very few of the females were then carrying eggs. A sample of 50 very small animals, collected at random, held only 1 (= 2%) ovigerous female. Samples from the northern bank of Lagoon C four days later on August 22, already showed 21% of the very small females carrying eggs; for the small animals this percentage was 38, for the medium sized animals 46 and for the larger females about 75%. Samples taken on August 29 showed percentages in the same order of magnitude.

It is assumed that in *Coenobita clypeatus* sexual maturity is not attained before the second year. Then many of the very small ovigerous females occupy *Tectarius* shells. Two samples of 50 very small animals each, all living in *Tectarius*, from a "playing ground" near Lagoon C showed gross weights ranging from 0.6 to 3.0 gr, with over 80% between 1.2 and 2.2 gr. In the meantime many much smaller specimens of *Coenobita*, not yet mature, are to be found. These miniatures have gross weights between 0.1 and 0.7 gr. Possibly they belong to an earlier year class, probably of the first year.

The number of eggs carried by these young females fluctuates around one thousand. Eggs of ten ovigerous animals were counted; numbers of 796, 859, 869, 871, 923, 957, 986, 1016, 1044 and 1260 were found. In the large animals these quantities may be 40 to 50 times larger.

Freshly spawned, still compact egg masses are dark reddish brown in colour. Gradually the eggs become paler, changing to grey or light blue, the structure of the egg masses becoming much looser.

From structure and colour of the egg masses a good impression may be obtained by the naked eye about the state of development of the eggs. Fully developed eggs, when placed in seawater, burst open immediately, releasing the young zoea larvae.

A few times females were observed, with sets of eggs of which the eggs on the frontal part of the abdomen had already fully developed and the eggs on the distal part were still brown and distinctly undeveloped. The gradual change in colour gives the impression that this may be caused by physiological factors, such as differences in temperature or oxygen supply, rather than that one or more new sets of eggs have been attached later on by the female.

Sets of eggs, attached to the female and enclosed by a proper shell are well protected against damage and desiccation. The relative humidity in the shells is supposed to be near a hundred percent; this, however, was not measured. There is always the moist skin of the abdomen and usually shell water is also present, but not in such quantities as observed in normal, non-ovigerous, animals. Ovigerous females in transparent artificial shells usually showed small amounts of water present in the apices of the shells, often moistening the eggs but never permanently bathing them.

In a small number of ovigerous females, caught on August 21 and 29 near Lagoon C, it was possible to shake out enough shell water for conductivity measurements in the laboratory. The salinity of this water proved to be quite normal, only a little below the salinity of 100% sea water (Table 8).

TABLE 8

SALINITY OF REMAINS OF SHELL WATER

measured in 7 ovigerous females from Lagoon C.

nr.	date, 1967	eggs	salinity ‰
1	21.VIII	freshly laid	34.4
2	21.VIII	freshly laid	35.0
3	21.VIII	freshly laid	31.3
4	29.VIII	half developed	35.0
5	29.VIII	half developed	32.1
6	29.VIII	half developed	35.8
7	29.VIII	half developed	33.5

It is quite comprehensible that in females, carrying nearly developed or fully developed eggs, there is no water in the shells, since absence of water prevents hatching of the eggs in the shell cavity. Only 4 out of a sample of 50 ovigerous females, caught southwards of the Landbridge on September 11 and all carrying mature eggs, showed traces of water in the shell, which may also have been some water from the branchial cavities.

Ovigerous females in captivity often lose the eggs prematurely, which could be observed in particular when females were placed into too small shells, deforming the loose clusters of developing eggs into compact masses.

Under circumstances as met with in Curaçao egg development in the field takes 3 tot 4 weeks. Interrupted periods of egg-production and also, to a certain extent, differences in physiological and ecological factors will play a role in the observed differences in the duration of maturing of the eggs. When clusters of three weeks old eggs are placed in sea water they yield only small percentages of swimming larvae. If the same is done with eggs which are a few days older, large quantities of young larvae are released.

Ovigerous females pass over the Landbridge only if nearly all the eggs are fully developed.

Determinations of sex ratios on the Landbridge as well as in the adjacent part of the coast gave an extremely low percentage of males in all size classes. The majority of the females carried eggs, smaller numbers were normal females without eggs and a small percentage of the females proved to be deficient. It is not quite clear why males should be absent and deficient females, along with egg-less females, should occur in smaller or larger numbers. Perhaps the distance between hiding places and beach are too short to permit a good separation. Moreover part of the animals here may be looking for sea water for drinking purposes.

Egg deposition

When strong movements towards the sea were observed on the Landbridge in the night of September 11, thousands of *Coenobitas* were already present on the shore, partially hiding under loose stones and in the shoreline, as well as active on the wet coral stone near the water level. The weather on that day was very quiet; the moon nearly in its first quarter and the tide was low. From the scattered clusters of eggs – rather inconspicuous on the buff under-ground – it was clear that egg deposition was going on. It was also clear that none of the animals was in direct contact, not even partially, with the water of the rock-pools or the open sea. A simple washing away of the ripe eggs (larvae) from the oostegites did not occur. For completeness' sake it may be mentioned here that fishermen on the island of Saba insist that the local soldier crabs migrate into the sea in September to "wash off the eggs" (Mr A. J. GEERLINK, pers. comm.).

Just as with the incorrect representation of the uptake of shell water, suggested to be a simple filling of the shell by the waves on the shore, the water management and salt regulation of *Coenobita clypeatus* does not allow such a coarse procedure. As a matter of fact complete inmersions caused by a sudden high wave or a fall may occur during the egg deposition, but always by accident.

What actually happens is that the female, with the help of the fifth pereiopodes, picks up small clusters of fully developed eggs from the masses attached to her abdominal appendages and takes them further out with the maxillipeds and chelipeds. In this way the eggs become spread over the wet stones and later on, during high tide, the larvae are washed into the ocean. In females hiding under stones, higher up on the shore, we often noticed that only part of the eggs were removed, and it is supposed that liberation of the larvae occurs on a number of consecutive nights, enabling the female to release only the completely developed eggs.

Several days after the females have deposited their egg-clusters, they are still recognizable by the whirled and coiled setae on the pleopodes and the yellow coloured remains of lute or fragments of the egg shells.

In order to learn more about the method of egg deposition considerable numbers of ovigerous females were brought to the laboratory. A few times it was possible to observe females kept in a wire netting cage, bringing out bits of the egg masses from the shell cavity. Then the eggs were twisted into a slightly spherical shape between the maxillipeds and afterwards attached to the tip of one of the two chelipeds. A short forward movement hurled the eggs away.

Unfortunately normal "egg-shooting" by females in artificial glass houses could not be observed, but it was sufficiently clear that the eggs are loosened from the setae of the pleopodes by the small scissors-like pincer of the left and right 5th pereiopod.

Some experiments described below were carried out to study the effect of the "egg-shooting." Glass aquaria of about 30 × 45 × 30 cm were divided into two compartments by a partition of steel gauze with a square mesh of 2 cm. At one side medium-sized females with fully developed eggs were placed and at the other side, 1 cm from the gauze a square plastic tray (20 × 20 cm) filled with normal sea water. Next day lots of swimming larvae were observed in the water, but also clusters of eggs beside the tray, attached to the gauze and to the walls and bottom of the animal compartment.

Evidently the eggs are not shot off in any definite direction and the water does not seem to have any special attraction.

More elucidating are the experiments in which the water of the tray was replaced by a layer of moist filter-paper, on which the eggs stay attached to the very place where they hit the paper (Fig. 6).

The maximum distance observed over which they were shot from a maximum elevation of about 25 cm, as was made possible in the experiments, amounted to about 20 cm. The numbers in Figure 7 refer to the numbers of eggs in each cluster on the paper. Again there is no question of a directional "shooting".

Yet this manner of egg deposition is quite functional for the animals, safeguarding them against entering the ocean. Theoretically *Coenobita clypeatus* would be able

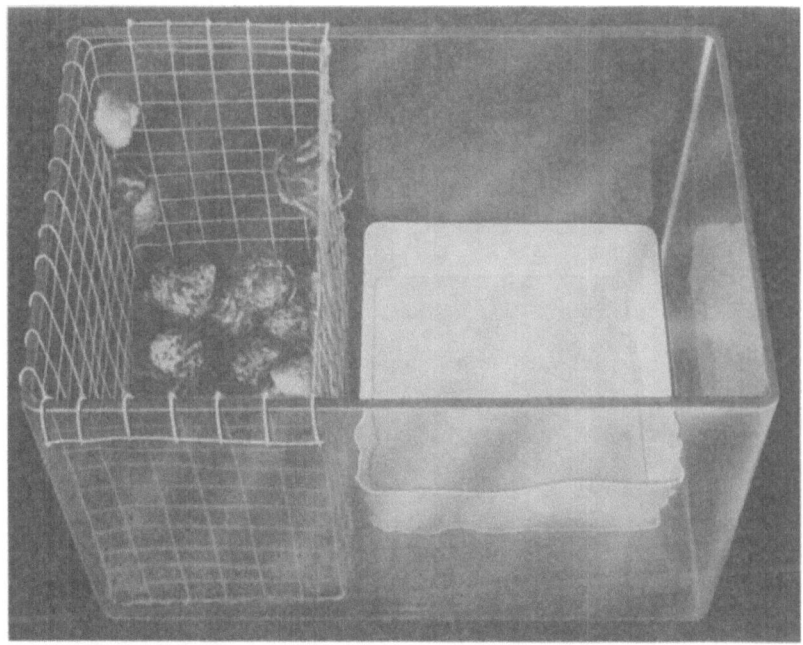

Fig. 6. EXPERIMENTAL SET UP FOR DEMONSTRATION OF 'EGG SHOOTING' by *Coenobita clypeatus*. — After females with fully developed eggs had been put into the left compartment, small clusters of larvae were observed on moist filter paper in the plastic tray to the right.

TABLE 9

LARVAL STAGES IN *Coenobita clypeatus* FROM LABORATORY REARINGS

after PROVENZANO, 1962. — Larval duration in nature may be shorter. — All stages are planctonic, except for the glaucothoe which shows partly a non-swimming behaviour.

stage	total length	duration
first zoea	2.6–2.9 mm	3–5 days
second zoea	3.2 mm	3–5 days
third zoea	3.8 mm	3–4 days
fourth zoea	4.6 mm	3–4 days
fifth zoea	4.7–5.5 mm	4–6 days
sixth zoea	—	4–6 days
glaucothoe	4.3 mm	over 1 month

46

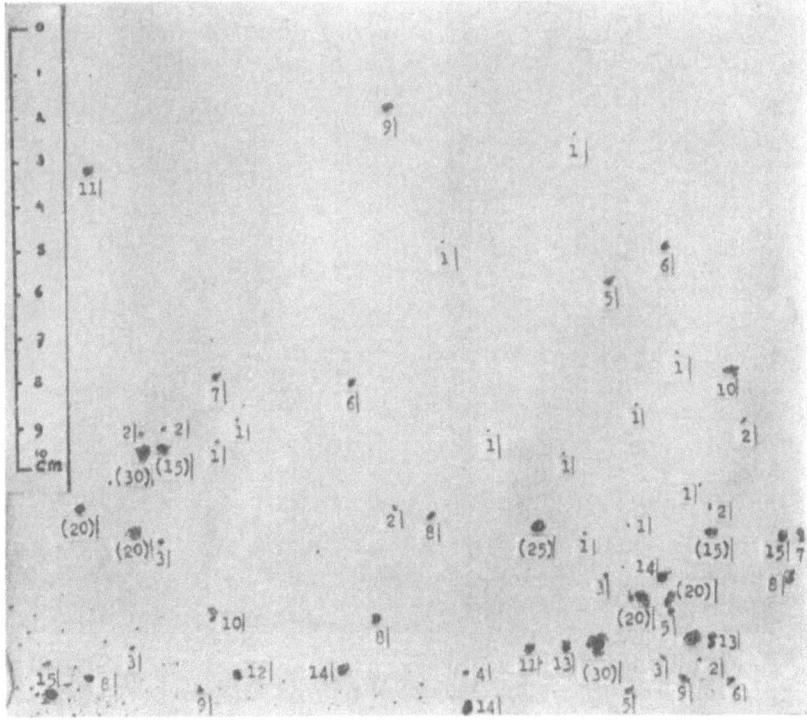

Fig. 7. EXAMPLE OF EGG SHOOTING RESULTS on wet paper. — The distance between the steel gauze (cf. Fig. 6) and the base line of the photograph was 2 cm. The numbers refer to the numbers of eggs in an egg cluster; estimated numbers in parentheses.

to drop the eggs from dry places above sea level, e.g. cliffs, etc., with fairly good results.

According to PROVENZANO (1962) larvae of *Coenobita clypeatus* hatch in a first zoeal stage, after which a number of three, four or five other zoeal stages are to be completed before they moult into the glaucothoe stage. Data from his laboratory rearings at a temperature of about 29°C and a salinity ranging from 33–36‰ are shown in Table 9.

A specimen which lived for 31 days as a glaucothoe and very closely approached the moult to the first crab stage, assumed a non-swimming behaviour about one week before it died.

Thus the total duration of the free-swimming larval stages in the laboratory covers periods from 40 to over 60 days. As the duration of the larval stages may be greatly influenced by diet as well as by temperature this period is probably shorter under natural conditions.

Successive reproduction periods

From the observations in 1966 and 1967 it was evident that the first group of animals, completing a sequence of mating – egg production – egg development – egg deposition, was followed by a partially overlapping second group of breeding animals, which in its turn was followed again by a third one. Studying the two observed periods in 1966 on the Directiestrand and three others in 1967 from both the Directiestrand and Oostpunt, a distinct lunar cycle is to be noticed. Figure 5 gives the range of each period and also the phases of the moon. Depending on the duration of the development of the eggs (3–4 weeks), the liberation of the larvae takes place in the week preceding full moon. In all cases the first young eggs were observed round about full moon. Egg production on the 2 places at Oostpunt and the Directiestrand seems to be synchronized. It is supposed that the introductory mating activities are primarily induced by the moon phase, roughly in the week before full moon. Correlations between moon phase and other stages of the reproduction are therefore secondary.

The numbers of animals observed always gave the impression that the first group of breeding animals was the most important one; the second and third groups of decreasing importance; however, no exact data about the first and second groups can be given. In any case the third group is very small in number. The observation of an ovigerous female in December 1966 is considered to be quite an exception.

Other records of consecutive breeding periods of *Coenobita clypeatus* are absent, but it is acceptable that this phenomenon will only occur in the central, warmer parts of its range, whereas in the northern parts, as Florida, the Bahamas, Bermuda, only one period occurs. Comparable data on other related species are also lacking.

Retreat to inland habitats

From the decreasing numbers of animals near the Lagoons around the middle of September, coinciding with the end of the first breeding period, it might be concluded that this first group disappears into the interior. It does not seem likely that females of this group stay and breed for a second time. Probably this holds true for males too. Sex ratios do not show a marked decrease of males. A number of the samples from the Lagoons, however, gave deviating sex ratios for the larger size classes in the period after about September 15, showing an increase of females or a relative decrease of males. This of course will be influenced in the first place by the direct depart of females towards the sea coast in the periods of egg deposition, which is also demonstrated in a very distinct way by comparing the sex ratios found on both sides of the Landbridge (Table 10).

TABLE 10

SEX RATIOS IN *Coenobita clypeatus* POPULATIONS
OF THE LANDBRIDGE, 1967

Ratio = number of males divided by number of females, per size class. Size classes: v.s. = very small, s. = small, m. = medium, l. + v.l. = large + very large. Average values computed from a great number of Oostpunt observations. Number of sexed individuals in italics.

	land side of Landbridge					
	27.IX	6.X	13.X	20.X	27.X	average values
v.s.	—	—	0.23 *23*	0.50 *2*	0.0 *2*	0.04–0.25
s.	0.11 *20*	1.14 *15*	0.0 *12*	0.11 *20*	0.20 *24*	0.25–0.75
m.	0.05 *20*	0.36 *15*	0.12 *29*	0.08 *13*	0.13 *26*	0.75–3.0
l. + v.l.	—	—	0.33 *4*	∞ *3*	3.0 *3*	3.0–∞

	sea side of Landbridge				
	27.IX	6.X	13.X	20.X	27.X
all classes together	—	0.0 *102*	0.0 *88*	—	—

About the retreat from the breeding places towards the inland habitats most details are absent. Probably this takes more time and goes on in a more dispersed and less directional fashion. There is no direct evidence that animals originating from Well 4, visiting the breeding place near the Lagoons, returned again to the same Well, though 18 tagged animals were observed here after the period coinciding with the breeding season, in which not a single one could be observed near Well 4.

Of the 304 orange (new) tagged animals, caught and released near the Lagoons, only 2 could be traced afterwards, again more westwards in a manchineel grove near Awa Blancu. Not a single of the orange or yellow animals released near the Lagoons was observed near Well 4.

In general the way back to the inland will be far more difficult to trace. The wet season has started now and rainfall in November and December in Curaçao is higher and more frequent in comparison to the beginning of the breeding season. Drinking places do not specially attract *Coenobita*, making observation not easy during that time.

Sooner or later the animals will return to their favourite inland habitats. As could be proved, even three years later tagged animals were still numerous in the hermit population near Well 4. According to observations by PEARSE (1929), however, there is apparently no evidence of a homing instinct directed towards fixed hiding places. Within a certain habitat an animal "beds down" in the nearest convenient shelter and returns to the same spot more than once by chance only.

II. ADDITIONAL NOTES ON THE
ECOLOGY OF COENOBITA

Land hermit crabs are animal species characteristic for coasts and islands in tropical and sometimes also in subtropical areas, the latter when washed ashore by warm currents. That they are still bound to the sea is primarily due to the fact that the fertilized eggs, at first attached to the female's abdomen, have to be brought to a marine environment to permit further larval development. Moreover, the sea is also the main provider of the larger gastropod shells, which serve as exchangeable houses. Another factor may be the presence of salt (chloride ion) in the environment (BOUSFIELD, 1968). In arid areas the often higher relative humidity of the coastal air may be of importance, while sea water of normal salinity always remains a safe and stable source of water, to fall back on in cases of emergency.

HABITAT

As to its habitat *Coenobita clypeatus* is not very particular. The areas concerned comprise such widely divergent sceneries as desert-like coasts, and shady parklands with shrubs and trees, sometimes up to a considerable altitude. The common feature in all these areas is that the sea is not farther away than 10 km, or 15 km at the utmost. Because of their way of locomotion land hermits prefer open grounds, preferably with a dry and hard soil; dense vegetations of herbs and grasses are rarely visited. Contrary to what might be expected hermit crabs avoid permanently humid areas, marshland, muddy banks of salinas and freshwater pools.

In Chapter I the characteristics of the area inhabited by *Coenobita* on the Leeward Group have been discussed in some detail. Of other places visited by the author especially Jamaica and Bimini must be mentioned because of their unusual character.

In J a m a i c a flourishing and numerous populations were found on the keys off Kingston Harbour. On Drunkenmans Cay, the size of which is probably not over one hectare, partly rocky with a vegetation of *Rhizophora*, partly a sand-bank, with a sparse vegetation of halophytes, many hundreds of medium-sized animals were found on the fallen leaves under thick mangroves.

During day-time the inactive animals had no further shelter. Some of them moved about or sat with the first antennae projecting. The peculiar thing is that the animals live and prosper in such a small area. There is no fresh water, except when it rains. The climate in this part of Jamaica is much the same as in Curaçao. Besides water taken up with the food – which is thought to consist mainly of edible matter from the shoreline – water is probably taken up from the moist humus as well. The moist micro-climate under the hothouse-like mangroves implies a low evaporation. The rest will have to come from the sea water.

Actually these animals should be classed as "coastal animals," however, the shabbiness, as known from the Curaçao animals, is lacking altogether. Large *Livona* shells can be found everywhere. There is no hinterland and therefore no inland migration. Since all reproductive activities must be performed along the restricted and isolated coastline this population would be admirably suited for further research on reproduction.

The numerous animals found on N o r t h B i m i n i in the monotonous Casuarine groves also merit closer attention. The Bimini group is situated on the western edge of the Great Bahama Bank, approximately 60 miles from the Florida coast. The climate is more temperate than that of Curaçao. Early November 1967, during the author's visit, it frequently rained during the night, with temperatures slightly over 20°C. By day it was sunny, with temperatures around 25°C. Smaller animals, mostly in white *Polinices* shells, lived in holes in the thick layer of fallen *Casuarina* needles. Larger animals practically all in possession of *Phalium*, *Tonna* and especially *Fasciolaria* shells led more nomadic lives and were even active during day time, in blazing sunlight. It could be concluded from the fact that unicellular green algae grew in the siphonal groove of the occupied shells, that this diurnal activity must have been going on over long periods. The animals were strongly attracted by a refuse dump of an adjoining hotel. Some small animals foraged in the branches of shrubs.

In the southern part of North Bimini the – mostly small – animals were to be found in the neglected gardens around the Lerner Marine Laboratory, both by night and day. At night considerable numbers of small animals were seen on the sandy beach at the western side of the island.

In Key Largo (Florida), Barbados, and Trinidad solitairy specimens of *Coenobita* were seen accidentally, but only during the night.

TEMPERATURE

The distribution of terrestrial decapods, and particularly of Coenobitidae, is restricted to areas with hot climates, thus reflecting their great dependence on temperature. Especially low temperatures limit their activity and determine their natural boundaries. Being poikilothermous animals, they more or less conform to the temperature of their surroundings, however, by choosing the correct micro-habitat and by a certain amount of individual regulation of temperature (BLISS, 1968) they can avoid extreme body temperatures for a certain time. The optimal temperature range for *Coenobita clypeatus* – probably between 22 and 34°C – is practically the same as the general range of temperatures in the whole Caribbean area, where those boundaries are only exeptionally exceeded. Extreme temperatures of 37°C and more, as recorded for the coastal areas of Venezuela and Colombia, seem less suitable. When such temperatures were offered in laboratory experiments a distinct decrease of activity was observed, which may also be connected with retention of water. Though it is not exactly known what the lethal temperatures are – which anyhow is only a doubtful concept without a time factor – it does not seem likely that in their natural habitats the animals are ever exposed to such temperatures. Their strong tendency to seek adequate shelter, together with their predominantly nocturnal habits protects them from extreme temperatures during the day. Therefore it is not necessary to consider the inevitable evaporation as having an important function in keeping the body temperature below a certain limit (BLISS, 1968). The lower temperature delimitation is found at 20–22°C, where all movements get much slower. Animals kept at 18°C for a longer time became more or less lethargic.

The northern boundary of the distribution of *Coenobita clypeatus* closely reflects the restriction of its chance of life at lower temperatures. On the North American continent the species is found from the east coast of Florida to about latitude 27°N (PROVENZANO, pers. comm.). In the more temperate Bahamas they occur farther to the north and in the even more oceanic Bermudas they reach their northernmost limit at 32°N. With average February temperatures of 17°C and lowest temperatures recorded of about 4°C the species here is at the margin of its existence. According to recent information (W. E. STERRER in litt. 1971; ABBOTT, 1972) *Coenobita* is now extremely rare on the Bermudas, possibly due to the absence of suitable big *Livona* shells. This may be doubted, however, since even specimens in very small shells reach sexual maturity.

The diurnal activity of *Coenobita* on Bimini may be explained by the considerable differences between day and night temperatures. If at night the temperature falls below the optimum range, while during the day it stays within that range, the normal night-and-day rhythm may be discontinued, humidity conditions permitting. If in the winter months the temperature during the day also remains too low, the animals do not come out at all. Good shelter offers the only opportunity for withstanding relatively low winter-temperatures.

According to WÜST (1969) in the warmest months (August–September) – the time when the larval development of *Coenobita clypeatus* is in progress – sea surface temperatures around 28°C are to be found all over the Caribbean. In the waters around the Bermudas, however, the temperatures remain some degrees lower. The somewhat earlier and shorter reproductive season in the more northern areas as compared to Curaçao is a question of temperature.

FOOD

Land hermit crabs are omnivorous animals feeding on all kinds of vegetable matter as well as on protein-rich food, especially on carrion. If they do not get the right kind of treatment in the laboratory, cannibalism may occur frequently. *Birgus latro*, the coconut- or robber crab, feeds especially on protein-rich *Cocos* and *Pandanus*-fruits (HARMS, 1932, 1937). In Curaçao the food of *Coenobita* comprises fruits, berries, seeds or germinating plants, whatever is to be found in the field. In such places sometimes large concentrations of animals are to be found, especially under the fruit trees in small orchards. All kinds of wild fruit, e.g. cactus-fruits or the "apples" of the manchineel tree, which are poisonous to most animals, are taken eagerly. In the dry season the disks of *Opuntia* and the trunks of *Cereus* are also gnawed, probably because of the moisture they contain. Newly-deposited droppings of horses and cows are taken as a source of food and water. Carrion is very attractive.

On Oostpunt a dead donkey provided food for hundreds of animals for weeks on end. Even when only an empty dried skin was left, animals were still entering it through the various holes. The dry skin functioned as a sounding box and the scuttling of all these animals could be heard over a great distance.

Coenobita is useful as a scavenger; not only on the waterfront but also around human habitations, refuse dumps, kitchen gutters, etc. According to local fishermen soldier crabs can best be lured by grounds of coffee.

WATER

In other chapters the role of both fresh and salt water in the environment will be discussed in detail. The occurrence of *Coenobita* in more inland habitats is closely connected with the availability of fresh or brackish water in these areas. Therefore it is the more remarkable that the animals are usually absent in places where there is plenty of fresh water. Even Poos Grandi and Poos Manzanilla, the two main rainwater basins on Oostpunt, filled with fresh water for the greater part of the year, do not seem to have any attraction at all. Neither near the other man-made reservoirs on Curaçao, where rainwater is retained behind dikes of earth, hermit crabs were practically ever found. They are completely absent in plains which are inundated in years with a high rainfall, where often luxurious swamp vegetations have developed, while those areas are still inhabited by *Cardisoma*. In the same way land hermits seem to be rare, or absent altogether, in any other area with a high rainfall. Fresh water seems to become attractive only if it is present in scarce quantities in an otherwise arid area. The water management of *Coenobita* is apparently based on an equilibrium maintained by, on the one hand, the pressure of a constant strong evaporation and on the other hand a subtle method of uptake from the environment (Chapters III and IV). Heavy rains may disturb the water regulation within a short time (Chapter IV). Water for the regulation of shell water is taken up in proportion to the salt content. The still hypothetical or partly revealed mechanisms for uptake of water from moist substrates (this paper; BLISS, 1968) or even from the atmosphere may be only an inconvenience to the animal in a wet environment.

It is generally accepted that the presence of salts has an essential function in active water transport. This might even be of still greater importance to a land hermit crab with its delicately balanced water management. Just as in our endurance tests with *Coenobita* (Chapter V), BLISS, WANG & MARTINEZ (1966) demonstrated for *Gecarcinus lateralis* that the animals remained in a better condition if the available drinking water contained a certain percentage of salt. The presence of both fresh water and a salt spray carried by the wind into the inland regions of Curaçao might create a favourable environment for *Coenobita*, similar to the situation suggested by BLISS (1968) for *Gecarcinus*. It is another example of the extent to which the animals depend on the sea even in inland regions.

III. WATER MANAGEMENT

INTRODUCTION

Even the most terrestrial species of decapods are still not completely independent of the sea and thus cannot be considered
completely successful land animals. It cannot be doubted that in the
general course of evolution the animals are only in a transitionary
stage and it is still an open question whether they will ever succeed
in obtaining a complete independence of their environment as far
as water is concerned.

It is certain that the direct step from sea to land is not an easy
one. One might expect that such a step would stand a better chance
of success in a moist climate. Still, terrestrial crabs are often typical
of tropical coastal areas in arid regions. Among terrestrial hermit
crabs *Coenobita scaevola* is a very common species of the arid coasts
of the Red Sea. *Coenobita clypeatus* is found especially on the arid
northern coast of the South American continent and a great many
islands in the Caribbean.

Terrestrial insects have solved the problems of living very
successfully in dry air, mainly by economizing their water management to a high degree. Evaporation is restricted to a minimum by
the complex construction of the tegument and by the existence of
adjustable spiracula. Water is imbibed by drinking or with the
food, but it may also be taken up from the air, or originate from
metabolism. Water loss through excretion is very small due to
rectal reabsorption (PROSSER & BROWN, 1962).

It is to be expected that a terrestrial crab, being originally an aquatic animal, will have problems in its water- and salt-management. It appears that for this group their success as terrestrial species depends on a great number of morphological, physiological and ecological adaptations as well as on a special behaviour, adapted to an economical use of water.

MATERIAL

For the greater part of the year *Coenobita clypeatus* is easily obtainable in Curaçao; only during moist weather conditions when the usual drinking spots are obviously of little attraction, the animals disperse in the fields and are hard to collect. During the reproductive season, ranging from July till October, most animals live in the neighbourhood of the breeding places along the coast.

Animals used for laboratory experiments were mainly collected in the surroundings of the old country house "Santa Catharina" on the north coast and in the yard of the country house "Klein Sint Joris" on Oostpunt. Both habitats are very similar. There is plenty of (fresh) drinking water and food. Moreover, in both places there are suitable hiding places and shady trees, including fruit trees. The greater part of the animals collected here belong to the type of inland animals; however, since the distance to the north coast is rather short, there are relatively high numbers of smaller individuals. Very large specimens are absent or rare, probably due to fishermen, who now and again collect hermit crabs for bait.

Unless stated otherwise only medium-sized animals were collected and used for laboratory experiments. Newly collected animals were placed either in wooden boxes with access to fresh or/and sea water, or in the outdoor crab pen. The latter was a 40 cm high, gauze-covered, concrete construction of 4 m², situated under the machineel trees in the grounds of the Institute. There were a 40 l water trough in one corner and against the two opposite walls long rows of hiding places constructed of concrete cable-tiles. A well-projecting iron roof, about 2 m high over the pen, gave shelter against rainfall.

In order to be informed on the main ecological conditions an electric temperature, as well as an electric humidity probe, were mounted in the pen and connected by cables with an automatic recorder in the laboratory.

Land hermit crabs survived very well in the pen, providing that sufficient food and water were supplied and, even at densities of several hundreds of medium-sized animals, mortality was found to be very low.

In most cases animals used in experiments were treated beforehand in various ways, in order to obtain a more uniform and properly adapted material. Depending on the way of treatment the following types of treated animals are to be distinguished:

normal-animals: animals which have had access to both fresh-(tap) and sea water for at least two weeks;

aqua destillata (A.D.) *animals*: animals throroughly rinsed and partially immersed in distilled water for 24 hours and afterwards kept in clean glass containers with access to distilled water only;

sea water (S.W.) *animals*: like A.D.-animals, but rinsed and immersed in 100% S.W. only;

desiccated (D.) *animals*: kept for at least 7 days without any access to water.

METHODS

Osmoconcentrations of fluids always were determined by the total number of dissolved particles, both electrolytes and non-electrolytes. Here the osmoconcentration of blood, urine and sometimes of shell water was measured by means of freezing-point determinations. For this purpose the micro-cryoscopic method and apparatus as described by HOHENDORF (1963) were used (Fig. 8).

The precision of the thermometer readings is about 0.005°C. In 1965, working at the Institut für Meereskunde in Kiel, with exactly the same apparatus as used by HOHENDORF, the same degree of exactitude as mentioned by him was reached by the author.

Blood was always sampled by means of a small glass capillary penetrating the membranous joints of the walking legs. Urine was sampled with a small curved pipette, which was brought into the urine bladder through the porus at the base

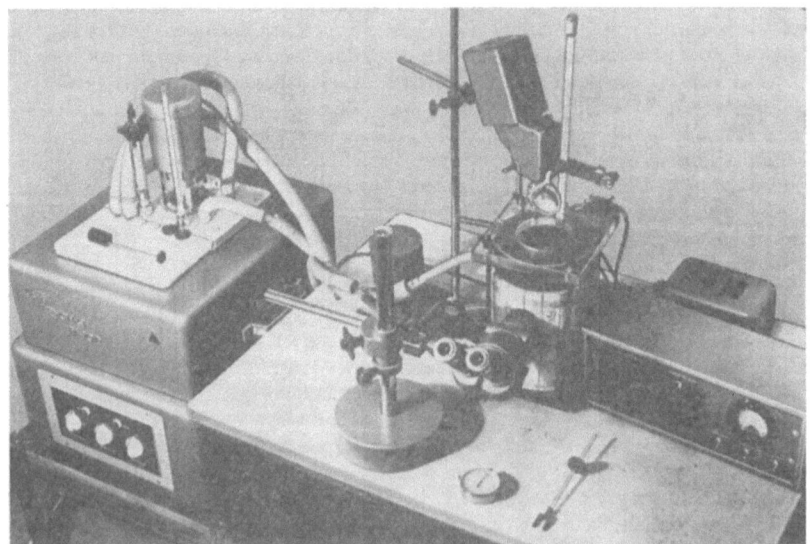

Fig. 8. DEVICE FOR MEASURING FREEZING POINT DEPRESSIONS. — Small samples of body fluids in glass capillaries are glued to a copper fork (front) and frozen in a Kryomat (left). Melting ice crystals are observed in a melting chamber by means of a microscope and melting points read on a thermometer.

of the second antennae. Neither blood nor urine were centrifuged before use, but as soon as possible they were prepared for cryoscopic determinations. As a rule the blood starts to clot soon after sampling and in 10 to 15 minutes it becomes gelatinous and therefore useless for proper determinations.

Theoretically samples of 0.3–3 γ are sufficient for one determination but in practice samples of 10–100 γ are more efficient for a proper filling of the capillaries.

Fresh samples of blood and urine were put immediately in watch glasses under a high quality liquid paraffin, after which as soon as possible glass capillaries with a diameter of 50–100 μ were filled with alternating drops of blood (or urine) and paraffin, in such a way that every capillary held about 10 drops from the same sample. A maximum of four capillaries, holding 4 different samples could be examined simultaneously. The samples were frozen in a "Lauda Tisch-Kryomat" type T.K. 30 D, at −36°C for a period of one hour. Melting ice crystals were observed in a melting chamber with a Zeiss Stereo Microscope III with a zoom system.

A very slow and gradual rise of temperature in the melting chamber is absolutely necessary for accurate readings. Therefore the rise in temperature was always fixed at 0.01°C per 1–2 minutes. Readings were made at the moment of disappearance of the last ice crystal of each drop. The average value of the first 5 or 6 fully melted drops per capillary was considered to be the freezing point depression of the sample.

For shell water and sometimes also for the larger quantates of blood – if concentrations of the total amount of electrically active particles were desired – determination of the electrical conductivity was performed in the same way as used by HOHENDORF (1963), with the only difference that a Philips Direct Reading Conductivity Meter, type P.R. 9501 was used. The micro-measuring cells permitted volumes of about 0.04–0.05 mm^3. Since, however, quantities 2 of 3 times as large were more satisfactory, the measuring chamber could first be rinsed with the liquid before measuring. All determinations were carried out at a constant temperature of 25°C; the results were recorded in milli-Siemens cm^{-1} and afterwards converted into salinities per thousand.

Blood was sampled in a similar way as described for cryoscopic determinations, but never stored under paraffin. Quick measurements are needed to prevent clotting in the measuring chamber.

Shell water was knocked out of the shells, caught in solid watchglasses and measured immediately afterwards, or – when sampling occurred in the field – stored in 2 ml glass jars with plastic caps and measured the next day.

For measuring drinking water and other water samples of a sufficient quantity, a Philips' Macro-measuring-cell, type P.R. 9513 was used, again at a measuring temperature of 25°C.

In order to be able to compare the results of both cryoscopic and conductivity measurements, most values were converted into salinities per thousand. To this end conversion-diagrams were made based on sea water dilutions and concentrations of known salinities by Mohr's titrations.

Beside the use of salinities, percentages of sea water (% S.W.) are often used to define concentrations of drinking water offered to *Coenobita*, in particular because the preparation of dilutions or concentrations of sea water is simple. The saline contents of the Caribbean sea water proved to be a little higher than "standard" sea water. (Compare I.A.P.O. Standard Sea water with a salinity of 34.95‰). In general the surface salinity of the sea water around Curaçao fluctuates between 35.80 and

36.80‰ (WÜST, 1964). In this paper 100% S.W. (normal sea water) is always based on a salinity of 36.00‰.

Temperature was usually determined by sensitive mercury thermometers up to a precision of 0.01°C.

If permanent readings were required two Philips miniature Pt. 100 Ohms resistance bulbs, type PZ, ± 0.1 at 0°C, connected with a Philips automatic recorder PR 3210 u/00, were at my disposal. If humidity and temperature were to be measured and recorded simultaneously, two nickel resistance thermometers Philips PR 6002 B were used (PHILIPS, 1963).

Permanent information about relative humidity of the air in experiments under laboratory conditions was obtained by using a combination of the aforementioned nickel resistance thermometers and Lithium-chloride Dew Point Hygrometers, Philips, type HA, both connected with the recorder.

In 1961 a Karl Weiss' Electric Hygrometer, "Hygronom," type nr 1381, was used for field work, whereas in some preference experiments in the humidity-apparatus dewpoint meters were used, according to the principle of the condensation hygrometer of Regnault, as designed by Professor A. PUNT of the University of Amsterdam.

For general information about climate and weather conditions in Curaçao the data of the Meteorological Service Netherlands Antilles, Dr. A. Plesman Airport, Curaçao, were consulted. The meteorological station is situated near the North coast of the island, about 7 km North of the Marine Biological Institute. Especially data on daily rainfall measured in a number of rain-gauges in the eastern part of Curaçao were most valuable.

Description of more specific apparatus and methods is given in the chapters in question.

GENERAL WATER ECONOMY

Only a few representatives of the class Crustacea have succeeded in solving the difficulties of a terrestrial life in a more or less satisfactory way. Most of these and related problems are compilated and summarized by EDNEY (1957, 1960).

Within the Decapoda only two groups show successful terrestrial forms: Land hermit-crabs of the Family Coenobitidae and True crabs of the Family Gecarcinidae. Semi-terrestrial forms within the Decapoda are much more common and a rather large number of species can stay for short times out of the water and show this behaviour in their normal way of life.

More or less semi-terrestrial species always keep water close at hand; either they can stay within a short distance from the shore, or they have burrows reaching down to the ground water. By immersing their bodies from time to time, at which occasion water may also very well be taken up, these animals avoid serious water problems (GROSS, 1955).

Genuine land crabs such as *Birgus latro*, various species of *Coenobita*, and *Gecarcinus lateralis* have become far more adapted to life on land (HARMS, 1929, 1932, 1937; GROSS, 1964; BLISS, 1968). They can stay for many days without taking up water and are able to roam far from their sources. This enables them to live nearly completely on fresh water. For the development of the larvae, however, they are still dependent on the sea and mature females have to go to the ocean to release their offspring. As is mentioned in this paper, it is not necessary for *Coenobita clypeatus* to enter the water for this purpose.

Water economy roughly involves three major parts: uptake, retaining and loss. Land crustaceans permanently suffer loss of water by evaporation. Provisions leading to economy in water consumption will be very important.

Uptake

Usually most water is taken up by drinking or with the food. Moreover, *Coenobita* is able to store water in its shell. This water may be used again partially as drinking water. Whether this water is also taken up by permeation through the soft tegument of the abdomen is not known.

Drinking and uptake of shell water in *Coenobita clypeatus* takes place in an active way. For other species another (passive) way of uptake of shell water is described.

By dipping the tips of the chelipeds, in particular of the minor right chela, in water, drops are lifted, handed over to the third maxillipeds and put into the mouth and gill cavity. Tufts of setae on chelae and maxillipeds play a role in this proces. This way of water uptake is very efficient and the slightest amounts of water e.g. moisture on surfaces of objects, occasional raindrops, etc. are utilized. Sometimes a more indirect way of drinking is found in which the tips of both chelipeds, held close together, are put into the water. By shoveling movements of the chelipeds, which bear setae on their innerside, capillary water rises to the mouthparts. Finally the maxillipeds bring the water into the mouth. VÖLKER (1965) described the same behaviour for *Coenobita scaevola*.

Coenobita clypeatus prefers drinking water of a rather low salinity but also water of higher concentration, up to a salinity of about 36‰ (= 100% S.W.) or even higher, is used. The higher concentrations, however, are unsuitable for use as shell water.

The way in which *Coenobita clypeatus* is able to trace water and the discrimination between water of different salinities points to a highly developed sensory perception.

Preference experiments

In order to obtain indications as to the way in which *Coenobita clypeatus* detects the presence of water and to which extent it discriminates between drinking water of different salinities, some preliminary preference experiments were carried out, which are briefly discussed below.

Soil humidity

Preference experiments on soil humidity were conducted in a narrow and rather high tank of a length of 220 cm, in which a slope of sieved coral sand had been set up. A considerable time after the sand had been moistened a tolerable gradient of soil humidity had established itself along the sand's surface. The process was checked by determining the wet and dry weights of sand samples. Along the sides of the tank alternating transverse partitions had been made, to provide the necessary shelter for the animals, in such a way that in this respect any part of the tank was just as attractive as the other parts. The whole apparatus could be shut off with shutters. Moreover an air current could be sent down the whole length of the sand's surface by turning on a small ventilator.

With the ventilator turned off regular counts of the number of animals in the respective compartments showed that they clearly preferred a zone at a distance of about $\frac{1}{4}$ from the lower end of the tank. The lowest part, where the sand felt wet, was avoided. The fact, however, that this observed orientation is shattered by turning on the ventilator, led to the conclusion that the animals do not use the soil humidity as such, but rather the gradient of humidity in the air just over the soil. With an air current this orientation is disturbed or prevented from establishing itself, so that orientation is made impossible.

Air humidity

The orientation on relative air humidity was tested in another apparatus. In an elongated deep tank bowls with evaporators as well as bowls with some odourless hygroscopic liquid (sulphuric acid) were put in such a way that a humidity gradient between about 50 and 100% R.H. was established. Over the bowls an experimental gutter was constructed from fine meshed wire, also provided with alternating transverse partitions. It was covered by a sheet of plateglass, with the dew-point meters screwed in at regular intervals. All experiments were carried out in the dark.

From the frequency distributions observed in the gutter compartments it could be concluded that without restriction the animals orientated themselves on the relative humidity of the air in such a way that they always went to the place with the highest humidity.

Salinity

In a third series of experiments it was tried to find out in how far *Coenobita* is able to discriminate between available drinking water of different salinities. To this end the basal part of the drinking apparatus described in Chapter IV was encircled by a round run of wire, divided by partitions into five compartments

TABLE 11

PREFERENCE EXPERIMENTS IN COENOBITA CLYPEATUS

Difference in salinity expressed in % sea water.
Differences in concentration in the order of 1 per cent are probably still perceived.
χ^2 are calculated from results in the columns pos. and neg.
Totals in italics.

difference in salinity	range	treatment of animals	number of animals	results pos.	neg.	indif.	χ^2
0.5% S.W.	0–0.5	dist. water	7	—	2	5	2.0
	5–5.5	dist. water	25	10	10	5	0
	99.5–100	dist. water	15	8	1	6	5.44
	99.5–100	sea water	16	8	8	—	0
	120–120.5	sea water	75	26	39	10	2.60
			138	*52*	*60*	*26*	*0.57*
1% S.W.	0–1	dist. water	11	4	3	4	0.14
	5–6	dist. water	35	20	9	6	4.17
	10–11	dist. water	31	17	8	6	3.24
	99–100	sea water	17 }	16	7	6	3.52
	99–100	desiccated	12 }				
	120–121	sea water	18	13	4	1	4.76
			124	*70*	*31*	*23*	*7.53*
2% S.W.	0–2	dist. water	23	11	6	6	1.47
	10–12	dist. water	3	2	1	—	0.33
	98–100	sea water	4	—	4	—	4.0
			30	*13*	*11*	*6*	*0.17*
3% S.W.	0–3	dist. water	11	7	3	1	1.60
	10–13	dist. water	3	3	—	—	3.0
	97–100	sea water	4	3	1	—	1.0
			18	*13*	*4*	*1*	*4.76*
4% S.W.	0–4	dist. water	12	6	3	3	1.0
5% S.W.	120–125	125% sea water	10	7	2	1	2.78
6% S.W.	0–6	dist. water	46	27	8	11	10.31

of equal size. Every compartment held two drinking places, connected with long and slender reservoir tubes. The water in the tubes differed in salinity. One animal was put into each compartment, and after a day the quantity of the water used from each of the reservoirs was measured. If the consumption from the two tubes differed less than 0.1 ml the animal was considered to have shown no preference.

In most cases the animals used in these experiments were given a preliminary treatment, to enforce a better choice, if the two salinities were nearly the same. One might assume that an A.D.-animal, or in other words an animal with a shortage of ions, will discriminate more acutely in the region of the lower salinities, 0–0,5% S.W., while e.g. a D.-animal or S.W.-animal profits most by a right choice in the region of the higher salinities, 120–120.5% S.W.

From the results obtained, which are best explained by table 11 one is led to conclude that differences in sea-water concentrations between $\frac{1}{2}$ and 1% sea-water, or in the order of 0.18–0.36‰ salinity may still be perceived.

Doubtlessly land hermits, which are omnivorous animals, will take considerable quantities of water with their food: fruit, succulent parts of plants, carrion, etc. It is evident, however, that these sources of water will be highly reduced in periods of drought.

As was observed in the field and subsequently also proved experimentally, limestone containing water may serve as a source of water in the dry season. By eating small pieces of limestone the animals are able to meet the need of water, to a certain extent.

Coenobita stores large amounts of orange-coloured fat or oily substances in its abdomen. Yet it seems that quantities of water formed in the process of fat metabolism will contribute only in a minor way to the water economy of the animals.

Retainment and loss

Both are closely related to each other, as a less economic management will lead to a more extensive loss of water.

Evaporation on the respiratory surfaces

Though the typical gills of the marine hermit crabs have undergone a number of modifications to make them more adapted to a terrestrial environment, such as a reduction of the number of gills, sclerotization, etc., still the respiratory surfaces have to be kept moist to enable a proper respiration. Moreover in this group of animals the respiration is highly supported by vascularisation of the walls of the gill cavities (HARMS, 1929; EDNEY, 1960). Air ventilation is supported by lifting the pleural margins of the carapace. Finally the most important feature in these animals is probably the skin respiration by the soft and moist tegument of the abdomen. Of course a proper gas exchange is possible only through more or less moist body surfaces; but this has the disadvantage of a serious water loss.

Adaptations to reduce evaporation on the respiratory surfaces appear in the shape of a reduced number of gills and also in a diminishing of the effective surface of the gills. The gill cavities are no longer filled with water, neither completely nor partially. Well-developed glands in the branchial cavity help to moisten the gills (HARMS, 1929). The gills act as a lung. The cavity has narrow connections with the air outside the animal. The moist abdomen is enveloped by a shell, the latter being carefully selected as to quality and shape (DE WILDE, unpubl.; VÖLKER, 1967).

Evaporation by the tegument

Evaporation through the exoskeleton of terrestrial crustaceans is highly affected by temperature, humidity and wind. It acts more or less as if it were a physical body (EDNEY, 1960). The tegument of terrestrial crustaceans, however, seems to be less permeable to water than that of non-terrestrial related species. According to HARMS (1932) the better calcification of the exoskeleton of *Birgus* should reduce the permeability. Layers of a waxlike substance, as are known in insects and spiders, do not occur (EDNEY, 1960).

Loss of water by excretion

Decapod crustaceans possess paired segmental excretory organs; the antennal or green glands. The excretion product is urine. End products of the nitrogen metabolism are discharged in this way and may play a role in the osmoregulation of the animals. The osmoconcentration of the thin, liquid urine of *Coenobita clypeatus* was found to be isosmotic with the haemolymph. Little is known about the quantities of water which are thus discharged. Loss of water by defaecation seems very low in times of water shortage. Faecal pellets are sometimes very dry and resorption of water in the end gut seems most likely.

Loss of water by ecdysis

Although loss of water during the moult will be considerable, this event, which only takes place at long intervals, needs not to be considered in the scope of this paper. The animals hide themselves as much as possible. The moult largely takes place inside the shell. It was observed how a moulting animal in a glass-blown shell managed to retain a certain amount of shell water in the shell during the moult. Recently moulted, still pale-pink coloured Coenobitas are always soft and moist to the touch and will desiccate very soon without the shelter of the shell and the hiding place.

Nocturnal activity

Besides direct provisions to reduce the loss of water a number of more or less secondary adaptations, which serve the water economy, must be mentioned here.

In the dry southern parts of the Caribbean *Coenobita* is almost completely nocturnal. At night the temperature is somewhat lower and consequently the relative humidity of the air is higher. The strong radiation of the sun is absent. Therefore the evaporation will be much lower at night. During daytime *Coenobita* stays motionless in its hiding place: in natural crevices, existing holes and tree stumps, under boulders, heaps of mould and decaying leaves, etc. Burrowing in loose earth as described for other species (VÖLKER, 1965) occurred in captivity, but was never observed in the natural habitats. Favourable places are sometimes inhabited by dozens of animals, staying in compact clusters. Such hiding places, though often found in very dry places, may show high humidities. In burrows of the terrestrial species *Gecarcinus*, BLISS (1956) found relative humidities of 95–100 per cent.

Provided land hermit crabs have a shell of the right size and the right shape, the animals are able to withdraw completely in the shell and close the aperture of the shell with chelipeds and legs. VÖLKER (1965) gives a detailed description of the closing mechanism as shown by various groups of hermit crabs, in particular of the terrestrial species.

In Curaçao some of the adaptations mentioned here or mechanisms concerning the water management of *Coenobita clypeatus* were further investigated.

Method

Coenobita clypeatus is considered to be nocturnal in its activity (PALMER, 1971). This was evident from own observations in the Leeward Group of the Netherlands Antilles as well as from the scarce notes on this subject by other authors, a.o. PEARSE (1916), CARPENTER & LOGAN (1945). HAZLETT (1966), who made observations on the social behaviour of the species in Curaçao says: "Tn Curaçao land hermit crabs were found at night, except for a few individuals found in the caves on the north side of the island."

Since a nocturnal activity involves certain advantages concerning the water management of terrestrial animals living in dry habitats in general, attempts were made to study the activities of *Coenobita clypeatus* in the laboratory. To this end the activity of about 200 tagged animals, living in the outdoor crab pen, was observed. The animals used had already spent ten days for acclimatization in the pen, prior to the first observations on April 23, 1967. Every day at about 17.00 h cooked rice, bread and fruit, and sometimes fish or other food rich in proteins was placed in the pen. The drinking trough held always fresh water. Obviously the animals were in optimal condition and mortality was observed to be negligible.

A temperature- and humidity probe, placed in the centre of the pen, about 20 cm over the bottom, and connected with the recorder, gave permanent information on temperature and humidity.

All visible animals in the pen were considered to be active; nonvisible specimens, e.g. animals hiding under the tiles, were considered to be at rest. This rather simple criterion as an indication for the activity of *Coenobita* was most useful in practice.

Counting the visible specimens takes only little time and therefore hardly disturbs active animals. In particular this was an advantage at night, when a small light source was used. Moreover, a small number of observations was made in order to specify the activity of the visible animals as to feeding, drinking, locomotion, attitude activity (the animal is nearly immobile, but the body is raised and protrudes partially out of the shell, often the antennulae are moving) and inactivity (the animal is drawn back in the shell and the shell is closed). These observations showed a very small percentage of inactive animals. Of course motionless animals which were observed to remain for a long time in the same place, were also considered to be at rest.

Additional inspections to observe the hidden animals in the compartments, by suddenly lifting the covering tiles, usually demonstrated inactive Coenobitas. Exceptionally active animals were also observed in the compartments, e.g. shortly before sunset or during sudden showers. In such cases, however, nearly all animals in a compartment were active, apparently ready to leave their hiding place.

Results and interpretation

The results of the described experiment are combined in Figure 9. Here the activity, as indicated by the total numbers of visible (active) animals and the relative humidity of the air in percentages are plotted against time. Night and day (from 6.15–18.45 h) and the occurrence of rainfall are given as well.

It was a favourable coincidence that during the course of the experiment weather conditions proved to be rather variable. Beside completely dry days – the normal type of weather in this month of the year – a number of short showers also occurred. April 26, and to a lesser extent May 5, were even wet days with drizzling rain over rather extended periods.

The activity curve shows a distinct rhythm; activity being high at night and low during the day. As an effect of the differences in temperature between day and night, the humidity curve in

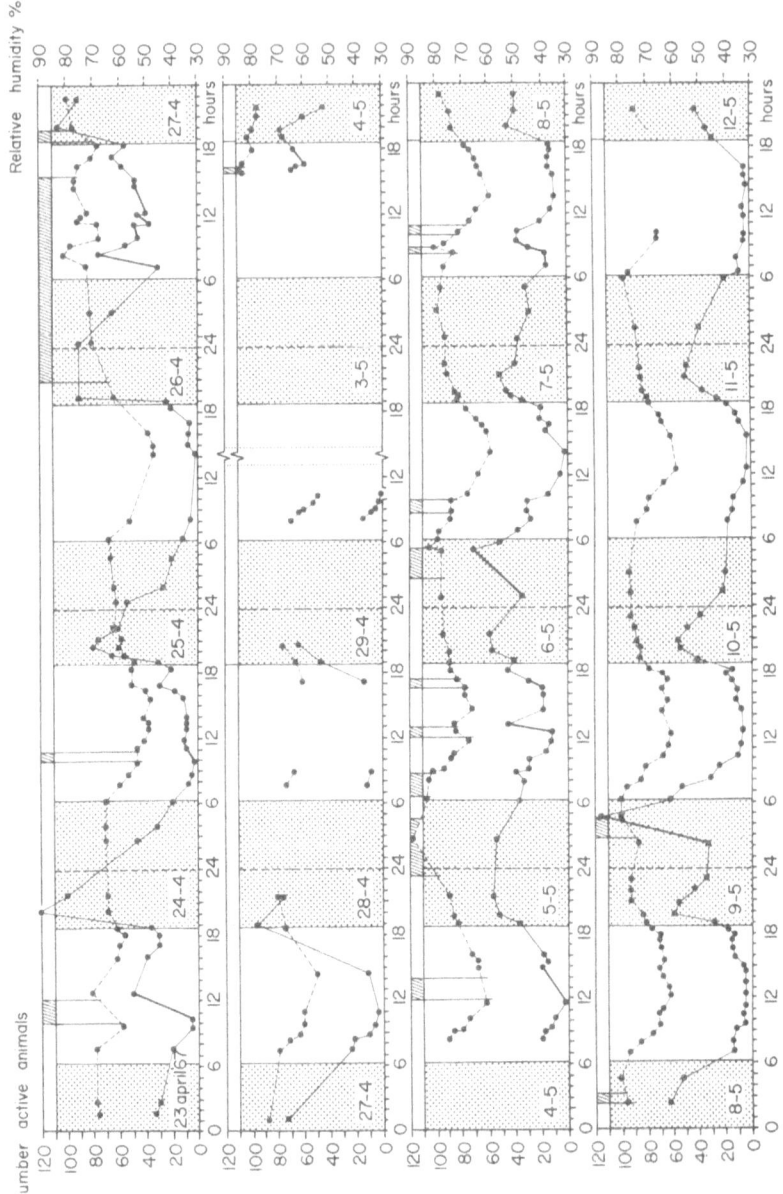

general gives the same daily fluctuations, i.e. lower R.H. during the day and higher humidity at night.

Presumably the basic pattern of the activity of *Coenobita* is based on a normal 24 hours cycle. PALMER (1971) found that nocturnalism in this species is under control of a biological clock. About noon the activity is lowest. Activity increases a little during the afternoon and shows a sudden and pronounced rise shortly after sunset (18.45 h). The highest activity is reached between 20.00 and 21.00 h. From now on little by little the activity decreases until the low values are reached about noon.

It was also obvious from field observations that activity of the species reaches its maximum in the first part of the night. Studying the activities of *Coenobita* on Oostpunt in Curaçao, during the reproductive season 1967, places along the water wells and lagoons were often found to abound with animals shortly after sunset, whereas some hours later all had disappeared.

Besides the basical day- and night-rhythm it may be expected that the relative humidity also affects the activity of *Coenobita*. In most cases a sudden rise of humidity brings on an obvious increase of activity. On April 26, a day with rainy weather and an almost constantly high humidity, the greater part of the activity-curve is shifted to a higher level. The same is seen on May 5 and again in the nights of May 6 and 9. Sudden and often considerable rises of humidity nearly always occur near sunset or as a result of rainfall. In both cases a fall in temperature occurs. A simple linear relation between activity and humidity, however, seems to be absent. Compare e.g. a relatively small rise in humidity in the night of May 8–9, effecting an enormous rise in activity with the effect of a much larger rise in humidity in the night of May 4–5. Apparently also other factors, perhaps the inclination or condition of the animals itself, will play an important part in this matter.

The secondary character of the humidity in respect to the diurnal

Fig. 9. PATTERNS OF CIRCADIAN ACTIVITY (solid line) expressed in numbers of active animals in relation to day and night rhythm and relative humidity of the air in percentages (broken line) during a number of days in April and May 1967. — Periods of rainfall (hatched columns in horizontal bar) generally induce sudden rises in activity (thickened parts of solid line).

rhythm is well shown by the steady decrease of the activity during the second part of the night, whereas on the contrary the relative humidity in general reaches its highest values shortly before sunrise.

Any direct influence on the activity caused by changes in temperature was not tested and is left out of consideration. However, if such an influence exists, the effect of temperature, as met with under natural conditions in Curaçao, will probably be small and the reverse of the effect of the relative humidity.

Hermit crabs kept under laboratory conditions at low temperatures, 15–18°C, showed very little activity; even in combination with high relative humidities. At a temperature of 18°C, for a longer time, they became more or less lethargic. As will be expected combinations of high temperatures (28–32°C) and high humidities (over 90%) induced high activity; high temperatures and low humidity (10–20% R.H.) minimal activity. Appropriate equipment enabling experiments at different temperatures under constant conditions as to the relative humidity was not available.

Limestone as a source of water

It is well known that terrestrial crustaceans are often found in the neighbourhood of caves, in particular in the limestone caves along the north coast of Curaçao.

Investigations in December 1960 in a number of caves and excavations in the steep escarpment of Rooi Rincón near Hato, Curaçao, did indeed reveal higher humidities (up to 95%) of the air, as compared with the much drier air outside (about 70% R.H.). There was, however, no water to be found, except for the moisture incorporated in the rocks. Of much more interest, however, were the formations of a very soft and porous lime-sinter stone, occurring in numerous places along the escarpment. Under conditions of sufficient daylight the limestone is locally covered by thick, velvety layers of green or blue algae, which doubtless indicate the high content of moisture of the stone. Wet and dry (24 hours at 110°C) samples of limestone, which were collected on December 8, 1960 during dry weather conditions, indicated weight percentages of water, present in the material, of at least 20–25%. As the attention of the terrestrial crabs appeared now to be focussed on these sinter formations, a series of laboratory experiments and observations was started to learn more about the advantages of limestone for hermit crabs.

Experiments and results

The experiments were carried out in two aquaria, each divided in two compartments by a double partition of small-meshed iron gauze and moreover provided with a grid of the same material, constructed about 3 cm over the bottom. In one of the two compartments of each aquarium a good-sized piece of limestone was placed, which beforehand had been carefully cleared of algae, whereas in both compartments D.-animals were housed; three experimental animals with access to limestone and three control animals without any access to that material, but staying under exactly the same conditions as to the other factors such as temperature (28°C) and humidity (90–100% R.H.). To prevent rapid desiccation of both animals and limestone as well as to guarantee similar humidity conditions in both compartments, a well fitting glass plate covered the whole aquarium. All the same it could not be avoided that the water content of the limestone decreased gradually

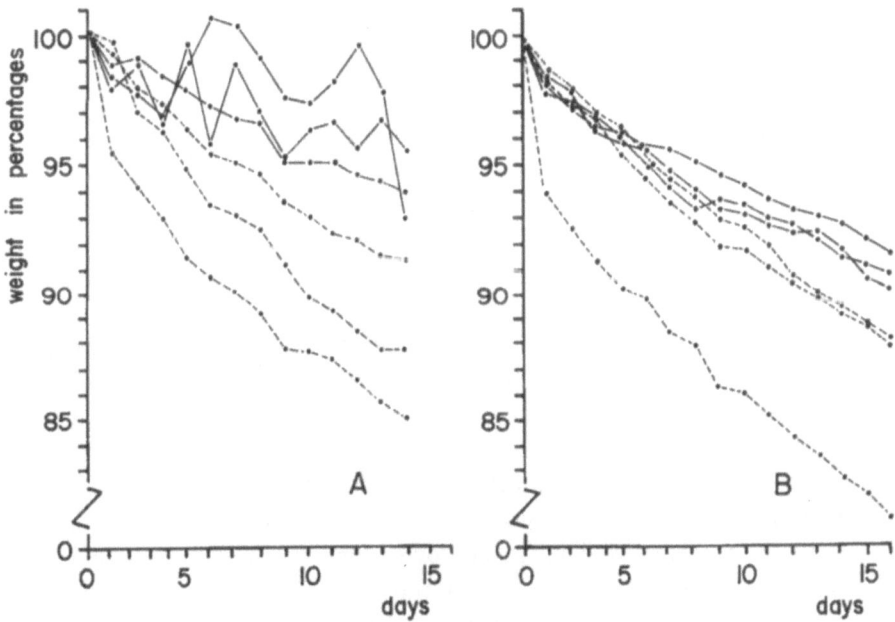

Fig. 10. LIMESTONE EXPERIMENT I. — Weight changes in percentages of 3 experimental animals (solid lines) and 3 controls (broken lines). A: experiment with soft porous limestone; B: experiment with hard crystalline limestone.

by desiccation to a value of 10–15% near the end of the experiment. Besides the experiment with soft and porous limesinter another experiment was set up in which hard and crystalline limestone with a water content of only 0.3–2% was offered to land hermit crabs. Changes in gross weights in all animals were checked by daily weighings. The first experiment lasted 14 days. Results are given in Fig. 10. Gross weights are converted into percentages of the net weight during the first day of the experiment.

Weight decrease in the control animals is gradual and well comparable in both aquaria. The graphs for the experimental animals with access to soft limestone represented in Fig. 10 A, shows obvious changes; total decrease in weight as compared to the controls is remarkably smaller and irregular, whereas distinct gains in weight occur now and again. In Fig. 10 B (hard limestone) differences in weight loss between experimental and control animals are shown to be relatively small. One animal shows a somewhat higher loss.

From the experiment the preliminary conclusion may be drawn that lime-sinter stone for some reason enables a better water management. Two more experiments, each with ten animals confirmed this conclusion. In these experiments the animals were seen to pick up small particles from the stone, using the smaller cheliped, and to transfer them to the mouth. This was observed at various occasions. In addition evidence was found that the lime-stone particles passed, to a certain extent, the alimentary tract, since the faeces of the experimental animals yielded ash contents between 85 and 90%, while the faeces of the controls had ash contents of some 50%.

In this connection the existence of a mechanism for water uptake, which Bliss (1963, 1968) mentions for the terrestrial crab *Gecarcinus lateralis* is interesting. Here the protruding pericardial sacs appear to transport water from moist substratum towards the branchial cavities, where it maintains a high relative humidity of the air and moistens the gills through direct contact between pericardial sacs and gills (Gruber & Shoup, 1969).

Convolutions of the membrane covering the pericardial sacs seem to reduce the surface tension of waterdrops which are in contact with the membrane. The water flows along the folds in the membrane and the posterior edges of the carapace into the branchial chambers. Abdominal setae along the posterior edge of the carapace facilitate the water movement (Bliss, 1968). Uptake of moisture from a damp substratum by means of tufts of setae between the second and third pairs of walking legs is also described for several species of *Ocypoda* (BLISS, 1968).

The shell as a protection against desiccation

Very young specimens of *Coenobita clypeatus* in the glaucothoe stage, when still living in the ocean, were already seen to search for a shell to protect the vulnerable abdomen (see also PROVENZANO, 1962). Immediately when going ashore the animals have the advantages of this shelter: protection against predators, protection against damage and chafing and, last but not least, protection against desiccation.

Though in the case of land hermit crabs the value of the shell for water conservation is obvious, as has been mentioned by a number of authors, the extent of this protection has never been examined.

In the next experiment the benefits of a well-fitting shell for *Coenobita* were proved by comparing desiccation in adequately housed animals and in the ones deprived from such housing. Twenty animals, freshly caught near Santa Catharina were used after all traces of shell water had been removed. Ten animals with a shell were compared with ten animals of which the shell had been removed by carefully cracking it in a vice. All animals were placed separately in small aquaria without access to food or water, exposed to temperatures of 28°C and a R.H. of about 75%, both quite normal conditions in Curaçao. Decrease in weight as found from daily or more frequent measurements was considered to be brought about mainly by loss of water. Loss of weight caused by defaecation proved to be negligible. Every series of weighings ended with the death of the animal. Finally, the empty shells were weighed to calculate the weight of the animal. For a proper comparison of the results net weights were converted into percentages of the net animal weight on the first day. Survival times proved to be extremely different for the two groups: 3–18 days, with an average of 7.6 days for the sheltered animals, and 1–2 days, with an average of about 30 hours, for the exposed ones. The rather large variations in survival time are due to the initial conditions of the animals, e.g. the uptake of water in time, the shape and size of the shell, the activity of the animal during the experiment.

It appears from this experiment that for properly equipped animals the time of survival is at least six times as long as for animals without shells. Moreover it will be clear that, in case shell water is present, the survival times will extend to several more days. To *Coenobita clypeatus*, living in a hot and rather dry climate, well-fitting and reliable shells are a matter of life and death.

The same experiment revealed that, for the majority of the animals in question, independent of their possession of a shell, desiccation of about 15–20% of the original net weight is lethal. Thus it is clear that survival in the first place is de-

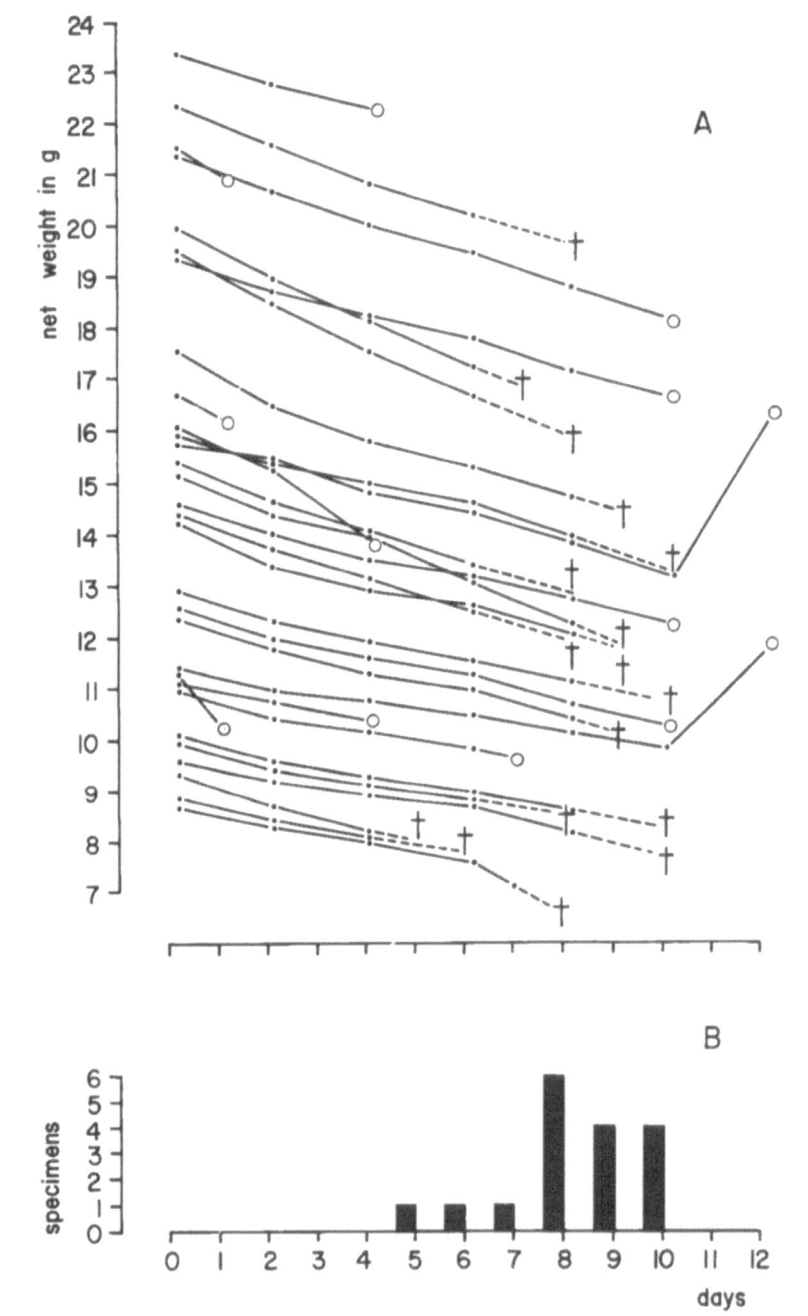

termined by the rate of desication, more so than by the time during which the animals are exposed to desiccation. As will be seen later, desiccation is always accompanied by an increase of the osmoconcentration of the body fluids and it is likely that when a critical value is exceeded, there will be increasing difficulties in the maintenance of vital biological processes.

Figure 11 represents the results of a second desiccation experiment, in which the changes in net weight in thirty, for the greater part medium sized animals, were measured under the same environmental conditions of 28°C and approx. 75% R.H. During the experiment, after 1, 4, 7 and 10 days, some animals, taken at random, were sacrificed for determination of the concentration of the body fluid. Two animals still surviving on the 10th day were replenished with fresh water, after which both the amount of retained water and the concentration of the body fluid were measured. For results concerning concentrations, however, the reader is referred to Chapter 5.

In an additional desiccation experiment with another 30 fairly large animals, some were removed every day and their loss in weight, as compared to the initial weight was determined. This was followed by determination of the dry weight (24 hours, 100°C). Thus the average daily loss in weight could be expressed in percentages of the total amount of free body water present in the animal (Fig. 12).

According to the experiments desiccation becomes lethal at weight losses of about 15 per cent or more of the initial body weight or of about 30 per cent of the initial body water. Environmental conditions as provided in the experiment bring on death in periods ranging from 5 to at least 10 to 12 days, with a rough average of 8 days. As the period of desiccation itself, as well as the loss in weight in percentage during this period are well comparable in both the largest and smallest of the tested animals, the actual loss of water in grammes per day must be smaller in the latter. The regression in the upper graphs (Fig. 11) is larger than in the lower graphs which represent animals of half the body weight of the former animals. In general the loss in weight in each animal is remarkably similar. During the first two days the evaporation may be somewhat higher, to decrease slightly in the next days and to increase again during the last few days before the animal expires.

As has already been described by other authors for a number of terrestrial brachyuran crabs, the almost gradual and constant

---

Fig. 11. A. WEIGHT LOSS BY DESICCATION in 30 specimens of *Coenobita*. — Relative air humidity about 75%; temperature about 28°C. O: used for conductivety determination of the blood; †: died of desiccation. On the 10th day of desiccation two animals were revived with water. — B. MORTALITY BY DESICCATION.

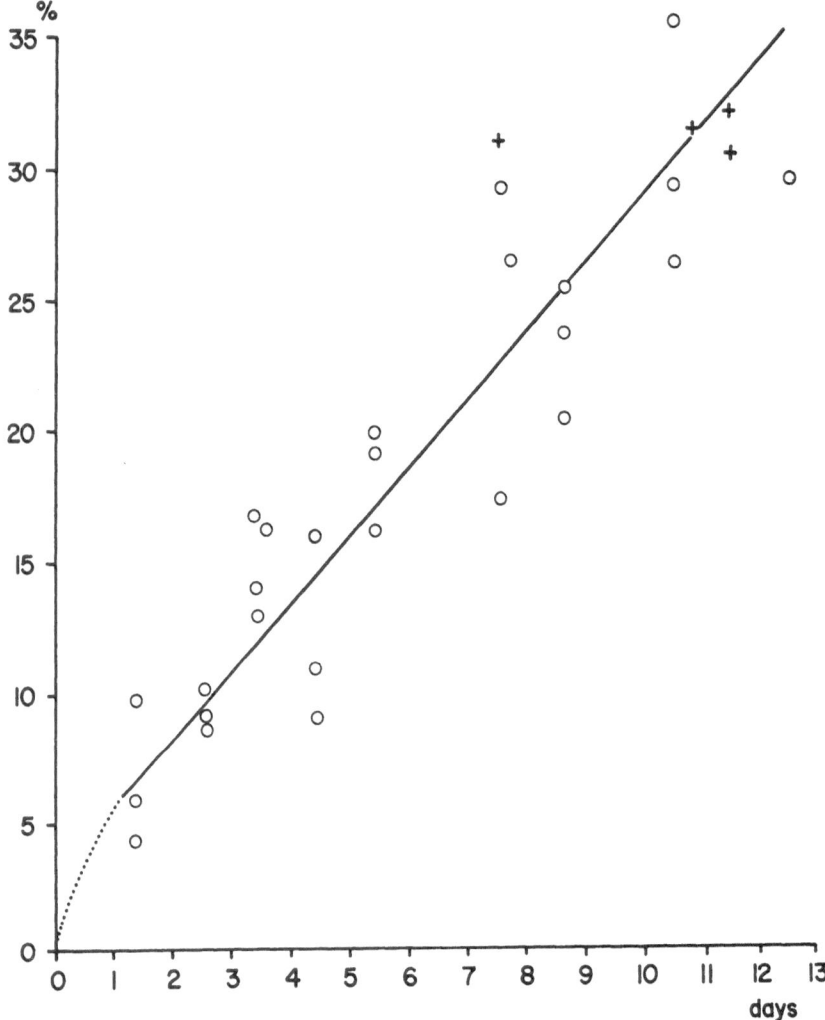

Fig. 12. WATER LOSS in percentages of total initial body water. — ○: weight determination in one specimen. †: died of desiccation. Dotted line showing influence of traces of shell water. R.H. 75%; 28°C.

decrease in weight in desiccating hermit crabs points to a passive way of dehydration: in other words to an evaporation similar to that of inanimate physical bodies (EDNEY, 1960).

In their normal life, however, land hermit crabs are enclosed by a more or less well-fitting shell, which may also contain a considerable amount of shell water, and therefore, though the principle of a fully passive dehydration for the animal itself may be justly applied, some complications in water loss by the animal in its shell are to be expected. Dependent on individual differences in size and shape of both the animal and its shell, the actual evaporation of each animal, kept under constant environmental conditions, may be expected to differ slightly, even when animals of equal net weight and equal activity are compared. To eliminate such variations in water loss as induced by animal-shell interaction, finally some desiccation experiments were carried out, in which the water loss under constant environmental conditions, during a certain period, was measured in several consecutive periods. In each series of measurements precautions were made to stop desiccation in time; then the animals were replenished with fresh water, after which they were exposed to desiccation again, and so on.

In general the presence or absence of shell water will represent a complication rather difficult to deal with; it seems very likely that the shell water will affect the pattern of desiccation. As soon as the animals are touched, they react by a deep retraction into the shell and in most cases this will cause a loss of water; the more shell water there was present originally, the more water is wasted. When, however, only moderate or small quantities of shell water are present, reliable results are possible, if the animals are held upside down during the procedure of weighing.

The great similarity in shape and gradient of the consecutive desiccation curves derived from a certain animal under equal conditions, is most striking (see Fig. 13). Dependent on environmental conditions the gradient of the curves changes. Besides the linearity a bend, occurring in the first or second day of each period of desiccation, is obvious. Most likely the latter represents the disappearance of the last remains of the shell water; or it may be caused by differences in the activity of the animal. From now on the water loss proceeds markedly slower. Linearity and equal regressions point to patterns of dehydration, well comparable to the evaporation of inanimate objects.

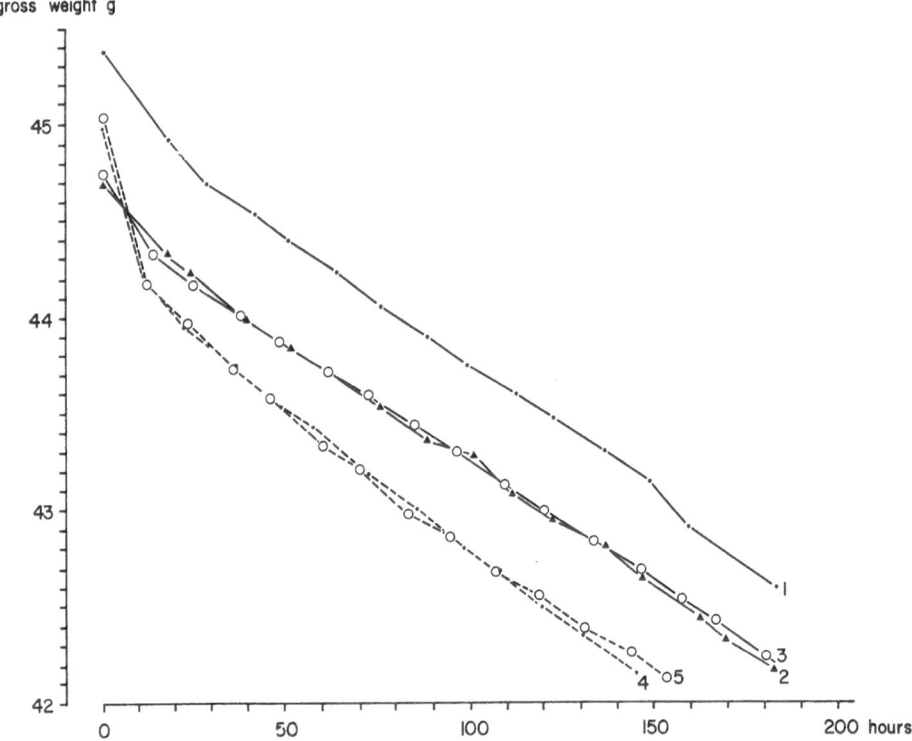

Fig. 13. EXAMPLE OF WATER LOSS by one *Coenobita*, in 5 consecutive periods of desiccation of about 8 days each. — Graphs 1, 2 and 3 at 22°C and 75% R.H.; graphs 4 and 5 at 28°C and 70% R.H.

DISCUSSION

Desiccation in semi-terrestrial crustaceans received fairly wide interest, cf. EDNEY, 1960 and BLISS, 1968. Comparable information on other decapod crustaceans deals mainly with brachyuran species. For *Gecarcinus lateralis*, one of the most purely terrestrial crabs, BLISS (1968, in Bimini, Bahamas, temp. 30°C, R.H. 78%) found survival periods of 89 hours on the average; 21% of the original weight was lost by the time of death. Similar data for *Cardisoma guanhumi* are 53 hours and 16% weight loss and for *Ocypode quadrata* 20 hours and 14% weight loss. Thus more terrestrial species can withstand a greater desiccation as compared to semi-terrestrial species, whereas the rate of water loss is lower. BLISS also found that all sizes of *Gecarcinus lateralis* are equally resistant to dehydration. In the other two species the smaller individuals proved to be more vulnerable. Evaporation of the exoskeleton in both terrestrial and semi-terrestrial crustaceans is highly influenced by temperature, humidity and wind (EDNEY, 1960). Waxlike layers, such as occur in insects and other land arthropods, have never been demonstrated in crustaceans, and water loss in terrestrial crustaceans always tends to be much greater as compared to insects.

For an appropriate discussion of the water loss in land hermit crabs, in the first place the biological observations and morphological facts observed by HARMS (1929, 1932) in *Birgus latro* and the Indo-Pacific species *Coenobita clupeatus* (= cf. *Coenobita brevimanus* Dana) and *C. cavipes* Stimpson, and in the second place, the more ecologically directed studies of VÖLKER (1965) on *Coenobita scaevola* are important.

HARMS (1929) emphasizes the great advantages for animals living in a shell, as regards protection against desiccation. The major chela and to a lesser extent the minor chela and partly also the flattened ambulatories together form a shield, similar to an operculum, which in well-fitting shells serves as an extremely useful closing mechanism (see for details on this subject VÖLKER, 1965). Those parts in the active animal protruding further outside the shell show increasingly thicker teguments and heavier chitinization. HARMS (1932) states that the importance of the abdominal respiration in *Coenobita* exceeds the function of the (modified) gills. The abdominal cuticles are very thin, showing, in particular along the dorsal side, invaginations resembling tracheae, and they appear to be moistened by special glands.

VÖLKER (1965) determined water loss and survival times in *Coenobita scaevola* under constant conditions in the laboratory. Three animals kept at 25°C and 75% R.H. survived for periods of about 9 days and on the average lost 18% of their original weight. At conditions of 25°C and 40% R.H. water loss proved to be $2\frac{1}{2}$ times higher. Survival times in two experiments, carried out at conditions of 29°C and 40 or 55% R.H. proved to be shorter (up to two days) in small animals (2–6 g net weight) as compared with periods ranging from 5–6 days in larger animals (8–12 g). The animals succumbed as weight losses by desiccation exceeded 20–30% of the original net weights. Moreover, VÖLKER calculated for his animals water loss in milligrammes per hour per gram animal weight and noticed, notwithstanding constant environmental conditions, that dehydration occurred irregulary. After initial high water loss an obvious slow-down was observed, which, however, did not tend to any constant value but went up and down alternatingly in a daily rhythm or with longer intervals. Finally, just before death, a final water loss of a greater volume occurred.

An explanation for these observations is not given. VÖLKER concludes that *Coenobita scaevola* does not possess any morphological structures to reduce dehydration. The only remaining possibilities are to close the shell and to become inactive.

From VÖLKER's experiments in *Coenobita scaevola* and our observations in *Coenobita clypeatus* it appears that the periods of survival may show rather large differences depending on: a) the strength of desiccation, b) the size and physiological conditions of the animals, c) the ability to close the shell adequately, d) the degree of activity of the animals during the experiment and e) the availability of shell water. Survival periods of one week or more are quite normal in both species of terrestrial hermits. In general these periods proved to be considerably longer than those found by BLISS (1968) for the genuine land crab *Gecarcinus lateralis* (e.g. 4 days at 30°C and 78% R.H.). Since Coenobitidae are able to stand periods of drought without any visible uptake of water, their success as land animals is apparent. Considering the dehydration which the species can stand during the desiccation, however, *Coenobita clypeatus* seems to take only a central place in the range of semi-terrestrial and terrestrial species. The extremely high values in *Coenobita scaevola* are remarkable (Table 12).

TABLE 12

WATER LOSS OF 'TERRESTRIAL' DECAPODS

expressed in percentages of original (net) weight at death.

species	%	author
Coenobita scaevola	25–30	VÖLKER, 1965
Gecarcinus lateralis	21	BLISS, 1968
Coenobita clypeatus	15–20	DE WILDE, this paper
Cardisoma guanhumi	16	BLISS, 1968
Ocypode quadrata	14	BLISS, 1968

Thus, when comparing *Coenobita clypeatus* and *Gecarcinus lateralis*, two of the most purely terrestrial West-Indian species, which very often occur in exactly the same habitats, the lethal desiccation appears to be be more or less similar. On the other hand the survival periods occurring during drought are twice as high in *Coenobita* as in *Gecarcinus*. It will be evident that the latter difference is mainly due to the adoption of a shell.

Another point worth discussing is the passive and uncontrolled water loss by semi-terrestrial and terrestrial crustaceans and by terrestrial hermit crabs in particular. If the teguments of these animals are indeed freely permeable to water and no regulating mechanism is present, the water loss may be better described as evaporation, rather than by the generally used term of transpiration, which suggest some form of an active process. Heavier chitinizations or calcification in terrestrial crabs may produce a decreased dehydration, but they do not provide a mechanism of regulation. Description of water loss in crustaceans as a pure physical process (EDNEY, 1960), observations by HERREID (cf. BLISS, 1968, p. 369) in dead (semi-)terrestrial crabs in which water loss continued at 75–95% of the rate at which they lost water when alive and the linearity of the desiccation curves in *Coenobita clypeatus*, all point to an evaporation process. Yet it is necessary to consider the question more closely. Although water is lost over the entire exoskeleton there is no reason to suspect that in semi-terrestrial or terrestrial crabs the respiratory membranes are particularly susceptible to dehydration (BLISS, 1968). To enable a proper functioning the membranes must be thin-walled in structure and moreover they ought to be kept moist. HARMS (1929) attributes the moistening of the respiratory membranes in *Coenobita* to special complex glands, which seem to be well-developed in the walls of the gill chambers, the head region and in certain appendages. In *Birgus* the secretion products of the "lungs" mix easily with water, forming a mucous fluid, which does not easily evaporate (HARMS, 1932). Microscopic slides of the abdominal tegument of *Coenobita rugosus* showed sub-tegumental glands, emptying into the transverse chitinized folds (HARMS, 1932). Although HARMS does not mention any function of these glands, a function concerning water secretion seems most likely. Assuming that HARMS' observations are right, it is well possible that there is some regulation in the loss of water induced by the activity of the glands.

Under constant environmental conditions weight loss by evaporation in a physical body is proportional to the surface of that body and further to the moisture content of that surface and the nature of the evaporating fluid. In living individuals the main source of evaporation – the stretched abdominal skin – shrinks in a very distinct way during desiccation; the often tightly stretched abdominal bladders fade completely away. Moreover the abdomen itself becomes drier, both at sight as well as to the touch, and finally the evaporating fluid continuously increases in concentration.

Thus a gradual decrease in the rate of water loss in desiccating hermits is to be expected. Although some of the curves tend to concavity, such a general decrease in water loss was not proved for *Coenobita clypeatus*, nor for *C. scaevola* in the study of VÖLKER.

This leads to the conclusion that the rate of water loss is mainly determined by the outer barrier, e.g. the shell and the "operculum."

Another anomaly in the desiccation pattern of *Coenobita* is the frequently observed rise in water loss some days before the death of the animals (Fig. 11, A). The same was observed by VÖLKER (1965). The suggestion is that the animals lose their ability to close the shell adequately. In addition there is a reason to suspect that the alternating periods of higher and lower water loss in *C. scaevola* are caused by periods of higher and lower activities. The effectivity of the closing mechanism is absent in periods of high activity.

Summarizing dehydration and patterns of water loss in terrestrial hermit crabs we may conclude that the shell ranks first in importance. Initial higher loss is due to the presence of shell water. Hereafter a remarkably constant water loss, mainly determined by the outer barrier (shell and operculum) occurs. A gradual decrease in water loss as is suspected by the shrinkage of the main evaporating surface, the abdomen, is not clear. Water secretion or secretion of mucus seems to be a possibility, but any special evidence for this assumption was not found. Variations in activity play an important role. Increasing water losses before the death of the animals are due to a less efficient closure of the shell.

IV. SHELL WATER AND ITS REGULATION

INTRODUCTION

It is a well-known fact that littoral and semi-terrestrial crustaceans, when staying ashore for a shorter or longer time, often carry some external water, usually retained in the branchial cavities (WOLVEKAMP & WATERMAN, 1960; EDNEY, 1960). The main purpose seems to be to moisten the respiratory surfaces and thus to enable breathing. Species such as *Grapsus*, *Ocypode* and *Uca*, as well as sometimes more aquatic species such as *Eriocheir*, vigorously aerate the water present in the branchial cavities. In more terrestrial species such as *Gecarcinus*, *Cardisoma* and *Coenobita* branchial water is not an absolute necessity; more or less moist respiratory surfaces seem to be sufficient.

The role of water in the branchial cavity in osmotic regulation, as discussed by VERWEY (1957) is probably of little significance.

In land hermit crabs of the genus *Coenobita* the dermal respiration of the soft and not-calcified skin of the abdomen has become very important and for this reason this part of the body must also be kept moist.

Land hermit crabs often store considerable amounts of water in the shells in which they live. Following GROSS (1964) this water is called shell water.

The occurrence of shell water in the Pacific species *Coenobita perlatus* (Edwards) was already noticed much earlier by SEURAT (1904), who connected its function with moistening of the gills. HARMS (1932) and VÖLKER (1965, 1967) mention shell water in two more species of *Coenobita*, pointing to the importance of keeping the abdomen moist.

In the case that rather large volumes of shell water are stored, which facilitate respiration as well as general water economy, enabling the animals to survive in dry inland habitats, its role in osmoregulation cannot be neglected. As far as known there is no information about the permeability for salt and water of the abdominal skin in land hermit crabs and although nothing is known concerning the osmoregulatory abilities of this part of the body wall either, one can imagine that such extremely thin teguments as described by HARMS (1932), continously bathed by shell water, will have some consequences to this effect. Shell water of a more or less constant and adequate salinity would be extremely valuable to the animals, but on the other hand this would imply some kind of mechanism to regulate the salinity of shell water.

Except for some observations by GROSS (1964) on the Pacific species *Coenobita perlatus* and *C. brevimanus* no attempts to study the concentration of shell water are known.

As recorded here, measurements of the concentration of shell water of *Coenobita clypeatus*, both in the laboratory and in the field in Curaçao, yielded salinities of a remarkable constancy, pointing to the existence of some mechanism of regulation. A number of factors of biological, physical and climatological nature, which were assumed to influence the stability of shell water, received special attention.

UPTAKE AND STORAGE OF SHELL WATER

Two different methods for taking up shell water may be distinguished, viz.: passive and active uptake. They depend on the hermit crab's way of life and may very well be characteristic for the species.

1. Passive uptake: The shell water enters passively into the shell. GROSS (1964) observed *Coenobita perlatus* H. Milne Edwards in Eniwetok atol, entering sea- and brackish water at night, which resulted in the adopted shells being filled with water. According to SEURAT (1904), who is also quoted here by WIENS (1962), *Coenobita perlatus* in Tahiti goes to the Lagoon shores every evening at sunset "to let the waves wet them and renew their supply of water." A similar behaviour was observed by VÖLKER (1965) for *Coenobita scaevola* Forskål in the Red Sea area. This species approached the waterline, so that it was wetted now and again by the waves. The animal squatted here for some time with the entrance of the shell facing the sea. After some minutes, when the shell was filled, the animals returned.

2. Active uptake: The active way of uptake is related with normal drinking, with the help of the chelae as described in Chapter III for *Coenobita clypeatus*.

Desiccated animals living in transparent glass shells (Pl. VIb) in the laboratory were placed on a grid of small meshed wire netting, some millimeters over a drinking bowl which was filled with red water, coloured by a tasteless dye. The animal could touch the water only with the smaller right-hand chela, so that every possible way of passive uptake by capillary rising was excluded. It was very clear that the glass blown shells were filled, using the active way of water uptake.

If an animal wants to store water in its shell, soon after the water is inspected and brought up to the mouth, red-coloured water is seen to flow into the branchial cavities. This observation is possible because the pleurae of the carapace are thin and slightly tranparent. Soon afterwards the animal raises the posterior margins of the carapace and the water runs backwards by capillary action and also helped by vigorous pulsating movements of the abdomen itself, filling the space between the inner wall of the shell and the abdomen and also the empty space in the shell behind the animal. Although the shells are sometimes full of shell water the animals walk about with the water without spilling any. In hermits in transparent shells it was observed that the water is not only held by capillary action but, also, that the skin of the anterior part of the abdomen is pressed firmly against the inner wall of the shell to prevent spilling. When, however, the animals are disturbed and have to withdraw deep into their shells a large portion of water is lost.

Concerning the water stored inside the animal's body attention has to be given to two bladder-like folds on the postero-ventral side of the abdomen. When drinking water of the right concentration abounds, these bladders become transparent after some time, bulging with body fluid.

To obtain information about the amount of retained water in eight Coenobitas the exact quantity was measured afterwards. For this purpose D.-animals were used, which had no access to water for 13 days previously and lived in prepared *Linona*-shells. These shells were sawn through longitudinally and fixed together with paraffin wax so that they could easily be opened at any time. The dry weight of the prepared shells (24 hours in exsiccator) was known. The gross weight of the desiccated animal was measured before bringing it to the water; afterwards the gross weight (now including the shell water) was measured again, and also the net weight of the carefully wiped hermit. Thus it was easy to find out which part of the water was taken up by the animal itself and which part was stored as shell water.

Figure 14 gives the results of the experiment. Net weights of the experimental animals in the last three days of a period of desiccation are shown. Fresh water is supplied on the 13th day. The gain in

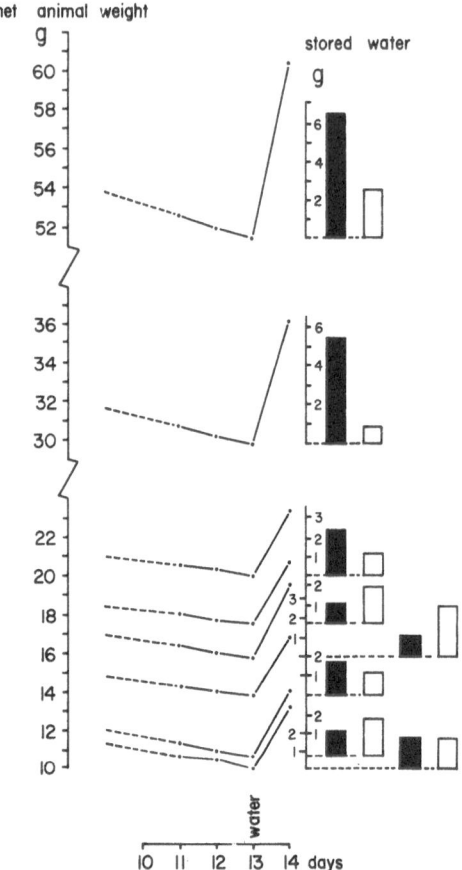

net animal weight

Fig. 14. UPTAKE OF FRESH WATER AND ITS STORAGE by 8 specimens of *Coenobita* after a 13 day's period of desiccation. — Dark columns: water stored in animal; white columns: shell water.

weight on the next day indicates the total of the water uptake for each animal, whereas the black and white columns represent the water penetrated into the body, respectively into the shell.

It is clear that considerable amounts of water are taken up here. At a later stage it was observed that in some cases the amounts of retained water might be much greater and that the amount of shell water alone could amount to a third or even half of the net body weight. Evidently the concentration of the offered water and the condition of the animals determine whether shell water will be retained and if so in which quantities.

An experiment as described above in which 100% S.W. or water with a somewhat higher concentration was offered also showed uptake of water, but now only in very small amounts. In these cases shell water is often not retained at all. Under natural conditions in the field it was also obvious, that animals depending on sea water store much smaller amounts of shell water than inland animals with access to fresh water.

From the observed differences in the stored quantities of shell water, depending on access either to fresh- or to sea water, it is obvious that *Coenobita clypeatus* discriminates between water of different salinities and moreover that water of a random salinity may be unfit to be used as shell water.

When, however, the concentrations of shell water were measured in a number of land hermits, which had only had access to fresh water, it was observed that in most cases the concentration of the shell water was much higher than that of the available drinking water.

This led to a series of laboratory experiments on the relations between drinking water and shell water. Essential in all these experiments was that water of various salt concentrations was offered to the animals, in such a way, that the volume of absorbed water could be exactly registered. For this purpose drinking fountains, as used for cage birds, were applied. In modified fountains with long tubular reservoirs, which were calibrated, water consumptions of 0.1 ml could be read off easily. Unless otherwise specified all experiments were performed under normal indoor conditions for Curaçao with temperatures between 26 and 32°C and a relative humidity between 70 and 80%. During the day the room was kept in twilight.

EXPERIMENTS

Uptake of water of a fixed concentration

Method

In a first experiment the total water consumption in a number of animals was measured when water of different salt concentrations was offered. The experimental set-up is shown in Fig. 15.

Fig. 15. EXPERIMENTAL SET-UP FOR MEASURING CONSUMPTION OF WATER of different salt concentrations. — The long calibrated reservoirs are fixed with plasticine on a plastic drinking trough. The first tank on the left does not contain any animals and is used for determination of normal evaporation.

In 7 glass tanks long tubular drinking fountains were put, holding water of respectively 0, 0, 10, 50, 100, 125 and 150% S.W. When the experiment started four A.D.-animals of equal size were placed in each tank, with the exception of tank nr I, which was used for measuring the normal evaporation of the cisterns.

The initial blood concentration of these A.D.-animals was known from determinations in other, completely comparable, animals. Every day the water consumption in the various tubes was checked by reading off the water levels; the reservoirs were replenished if necessary.

Except for 2 animals in tank II, with 0% S.W. and for the 4 animals in tank VII, with 150% S.W., all other animals survived for 50 days, the duration of the experiment.

Concentration of both shell water (by conductivity) and blood (by conductivity and freezing point determination) was measured at the end of the experiment.

Results and interpretation

For 5 successive periods of ten days each, the water consumption, expressed in ml per animal, is calculated. A correction for the evaporation in the drinking fountains has already been made.

Looking at Fig. 16, first period, the hermits in all tanks have a rather moderate consumption; nevertheless a distinct preference for 100% S.W. is shown, whereas distilled water seems to be the least suitable. During the second and following periods the absorbed volumes of 0, 10 and 150% S.W. still remain small, but now the uptake of 50% S.W. has increased markedly, whereas uptake of 100 and 125% S.W. has increased to extraordinarily high levels.

At first sight these results are in flat contradiction with the previous observations in which was stated that animals, having access to fresh water, always store large amounts of shell water, while on the other hand, shell water is scarce or absent in animals which drink 100% S.W. or water of a higher salinity. However, considering the behaviour of the animals, the observed differences are easily comprehensible. A.D.-animals encounter two major problems, which are the adjustment of the low concentration of the body fluids and the filling-up of the unavoidable loss of water by evaporation.

The data of the first period show that the animals in tank II only take very small amounts to avoid further desiccation but do not store any quantity of shell water, which would bring about further ion loss. Doubtlessly these animals are in very difficult circumstances, which is also demonstrated by the death of an animal in the first period and another in the second.

The same holds true more or less for tanks III and IV. As will be demonstrated below, the concentration of 100% S.W. is only slightly higher than the optimal concentration of the shell water and therefore animals in tank V are not only able to bring up the salt concentration of their body fluids to a normal level, but will also store certain amounts as shell water.

Animals in tanks VI and VII take some water to prevent further desiccation and adjust the concentration of the body fluids at the same time; uptake of shell water, however, does not take place, for the concentration of the available drinking water is far higher than the optimal concentration of the shell water.

In the second period the overall picture of the water uptake has changed. For the animals in tanks II and III the situation seems to be similar; at 0% S.W. another hermit crab has died, whereas the amounts of absorbed water are still very small.

In spite of the small amounts here and also in tank IV salt accumulation takes place caused by a repeated uptake and evaporation of small quantities of water of low salinity. Of course this will produce the smallest effect and will take the longest time in the case of 0% S.W. It was often observed how in these circumstances, Coenobitas stimulate a good evaporation by going abnormally far out

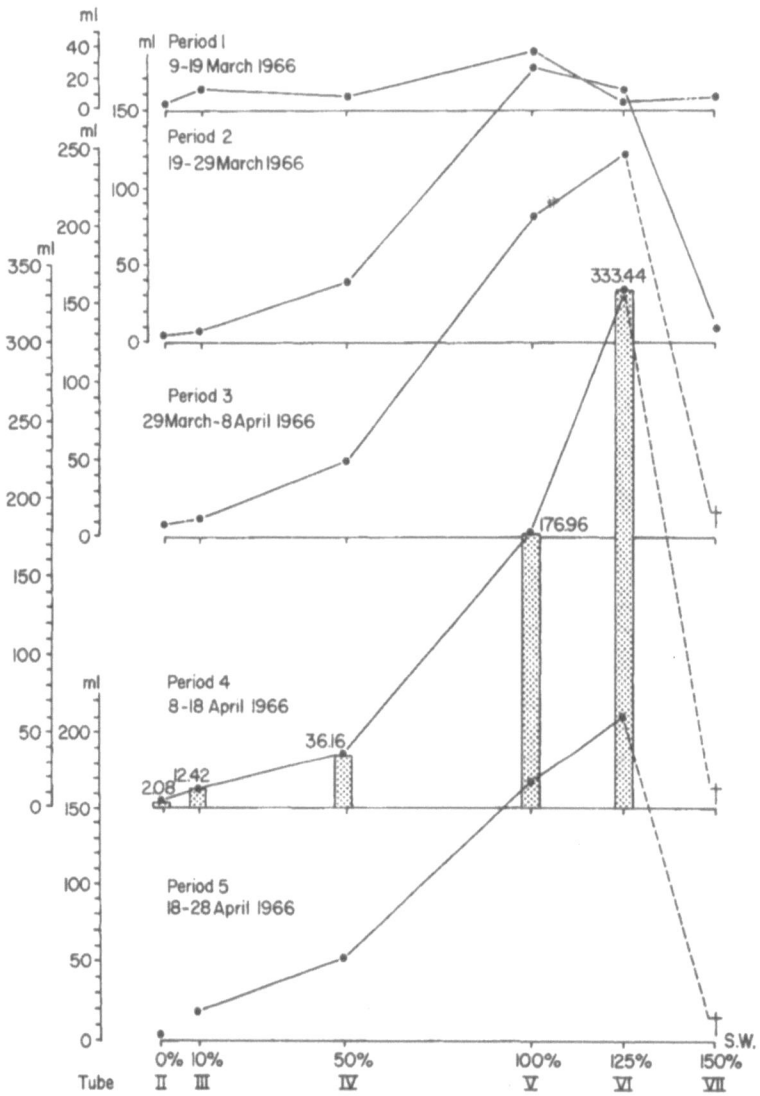

Fig. 16. CORRECTED USE OF WATER PER ANIMAL in 10 days, in ml. — Offered are concentrations of 0, 10, 50, 100, 125 and 150% sea water. In period 4 the consumption of 125% S.W. is more than 150 times that of 0% S.W. Columns drawn only in this period. 150% S.W. is not suitable as drinking water.

of the shell and sitting with legs extended on a well exposed spot, e.g. against the wire netting covering the tanks.

Considering the increased uptake of water of the animals in tank IV it seems most likely that these animals already have the right kind of shell water at their disposal.

Animals in tanks V and VI with respectively 100% S.W. and 125% S.W. behave quite differently. Such animals are also confronted with the steady increase of the concentration of the body fluids and – if present – of the shell water. When the concentrations of the body fluid or the shell water surpass the permissible levels, the animals start drinking again. From now on, however, they do not retain any quantity of this water, which would only add more salt to the animals, but they let it circulate through the gill cavities and the shell, while bathing the abdomen. After a short time the water flows out of the shell and the animal starts "drinking" again. In this way they spend many hours a day around the water fountains, pushing each other away, in order to replenish the old shell water and to keep the salt concentration of the body fluids as low as possible. The used volumes of water as read on the calibrated water reservoirs proved to be unexpectedly high.

Water of 150% S.W., as offered in tank VII is too salt to be used in a circulation as described. Only very small quantities of this water are used. One animal from tank VII did not survive the second period.

The strikingly different behaviour observed in animals which have access to drinking water of low salinity at one side and high salinities at the other, was also observed in the three other periods. Except for tank VII, in which the remaining animals all died during the third period indicating the total unsuitability of 150% S.W., no further mortality occurred. Obviously the animals of tank II had now passed the critical period.

At the end of the fifth period the remaining 18 animals were checked as to the presence of shell water and its concentration was measured. Both animals of tank II had very small quantities of a moderate salinity; animals in tanks III and IV proved to have large amounts of shell water of normal concentration. Although actually shell water was absent in the animals in tanks V and VI, in a number of cases it was possible to determine the concentration of the circulating water dripping out of the shells. At 100% S.W. the same concentrations as in the available drinking water was measured; at 125% S.W. the concentration of shell water was considerably higher. In this way the remarkable S-shaped regulation curve (Fig. 17) of shell water against available drinking water is found. Data on blood concentrations are incorporated in Fig. 32.

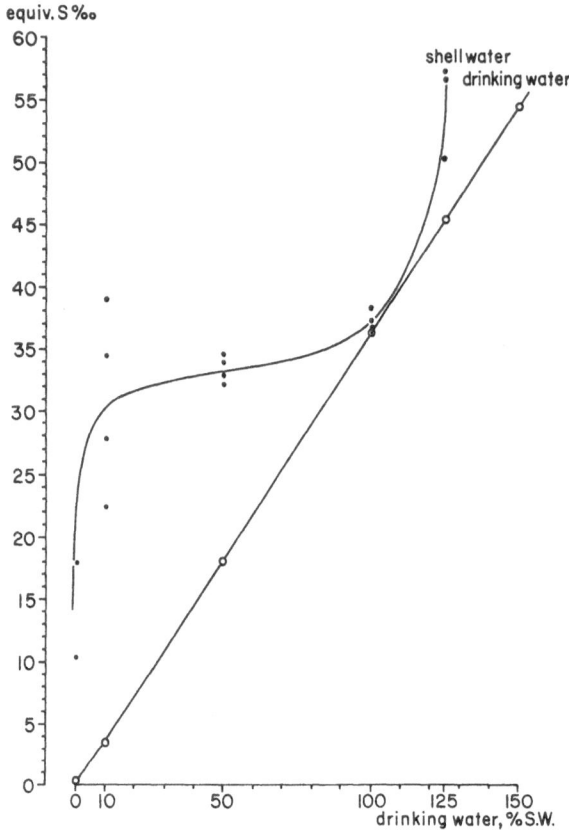

Fig. 17. SALINTY OF SHELL water in relation to the concentration of the available drinking water. (100% S.W. = 36‰ S.).

Without entering into details two important principles playing a role in the regulation of the salt concentration of shell water in *Coenobita clypeatus* are introduced here:

1) Accumulation of salts by repeated uptake and subsequent evaporation of small amounts of drinking water of low concentrations.

2) Circulation of large amounts of water in animals having access only to water of a relatively high concentration.

The results of the experiment mentioned above reflect the main difference between coastal and inland animals. The amount of absorbed water and the time needed for its uptake in coastal animals are much more than those on the inland animals, which use fresh water.

On account of their unsatiable need of circulating water, coastal animals are compelled to stay in the coastal regions.

Uptake of water in the drinking apparatus

Method

A second series of experiments describes hermits which are placed in the drinking apparatus. In this apparatus the animals can choose between sea water of different concentrations; from the absorbed quantities it is possible to conclude which concentration is the most attractive for *Coenobita*.

The apparatus consisted of a wooden standard, around which 10 drinking fountains, provided with long and slender reservoirs, were placed in a circle. The standard was enclosed by a cupboard with a door in front and a perforated bottom and top to permit a good ventilation. By revolving the standard the position of the fountains and reservoirs, which were fastened on the standard by means of tool-clips, could be changed; this in order to eliminate a preference for a particular direction and place. The experiments were performed under the same indoor conditions as described for experiment 1. The fountains were filled with 0, 2, 10, 25, 50, 75, 90, 100, 110 and 125% S.W. Thirty A.D.-animals were released in the cupboard. The experiment lasted for 40 days, or for 8 periods each of 5 days. The water available to the animals was constantly renewed, but measurements were only made in the 1st, 2nd, 4th, 6th and 8th periods. The experiment has been repeated a second time under the same conditions, with fresh animals.

Results and interpretation

The picture of the water consumption in both experiments is roughly the same: During the first period the A.D.-animals, which have a serious salt-deficiency, do not greatly discriminate in the uptake of water. Different concentrations in a fairly wide range are absorbed; however, moderate concentrations of 25, 50 and 75% S.W. seem to be most favoured, but a distinct preference is still lacking. Concentrations over 100% S.W. are seen to be less favoured. (The low amounts of 10% S.W. in particular when compared with the used quantities of 0 and 2% S.W. are difficult to explain).

During the second period a more distinct preference for 50% S.W. is obvious. In the next period a gradual shift to the use of ever fresher water is shown; finally resulting in a distinct preference for water of the lowest concentrations. In Figure

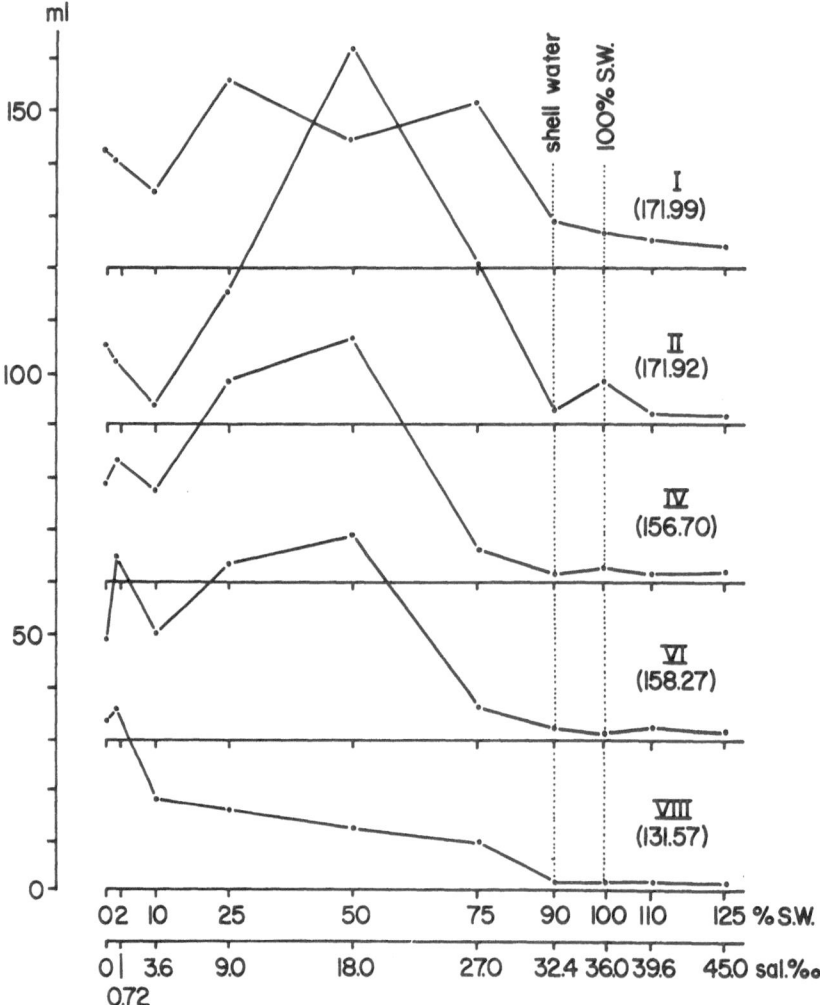

Fig. 18. USE OF WATER in ml per period of 5 days by 30 animals in drinking apparatus with 10 fountains filled resp. with 0, 2, 10, 25, 50, 75, 90, 100, 110 and 125% sea
water. — Dotted lines: concentrations of normal shell water and normal sea water;
between parentheses: total amounts of water absorbed in ml per period.

18, giving the results of one of the experiments, the consumption of water is plotted
for periods of 5 days.

A check on the presence and concentration of shell water, carried out at the end
of the experiment, indicated that 28 out of 30 animals had considerable amounts
of water in their shells. The abdomens and insides of the shells of the two remaining

animals proved to be only slightly moist. In 26 cases the concentration of shell water could be measured; see Fig. 19.

Though the animals unmistakably prefer drinking water of low concentrations, the concentration of their shell water is still maintained at much higher values, which correspond to 75–100% S.W.

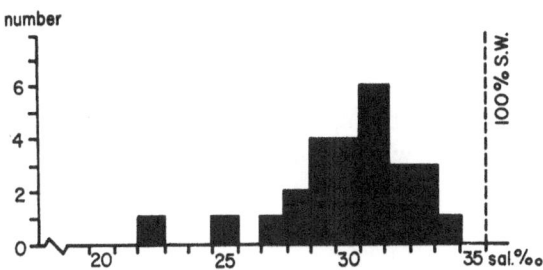

Fig. 19. DISTRIBUTION OF SALINITY OF THE SHELL WATER of 26 animals from experiment 2. — Broken line: 100% sea water.

Thus *Coenobita* takes up water of low concentrations in order to keep the concentration of the retained shell water on a more or less constant and much higher level. In other words: the shell water, which by evaporation grows salter continuously, is kept on a correct level by dilution with fresh water.

It is assumed that as soon an A.D.-animal is released in the drinking apparatus it will start to adjust its salt balance, either by taking up water of the right concentration or by a combined absorption of water of high and low salinity in the right proportions. In addition an extra supply of water of the right concentration will be retained in the form of shell water. Since its concentration will now increase by evaporation, the animal takes up a certain amount of fresh water for dilution. The fresher the absorbed water, the smaller the volume which will be necessary to obtain the desired effect. The uptake of water with higher concentrations (over 25% S.W.) shows a very distinct decrease during the successive periods, which on the other hand is compensated by the increased uptake of water of 10, 2, and also 0% S.W.

Moreover it is seen how during successive periods, the total amounts of absorbed water decrease from about 172 ml in the first period to 131.5 ml at the end of the experiment.

Considering the results of the previous experiments, it is interesting to speculate upon the process of passing from a marine to a terrestrial environment by coastal animals. Generally speaking any establishment in a terrestrial habitat will involve a passage from salter to fresher water and moreover this water will not be as easily available as sea water is. Such animals have not only lost the necessity to have sea water circulating continuously, but, as the water becomes less salty, they can also manage with ever smaller amounts of drinking water. In this way they become less dependent on water, which implies a higher degree of terrestrial adjustment.

Regulation by mixing of 0 and 100% sea water

Introduction and method

After the existence of various ways of regulation of shell water in *Coenobita* had been recognized, the next step was to investigate the in- or external factors, which might influence the process of regulation, such as the actual limits between which the concentration of shell water is kept, and whether the existence of such an optimal concentration could be proved.

To this end it was assumed that a simple device as shown in Fig. 20, inset, representing a tank with two drinking fountains, filled with 0 and 100% S.W. respectively, would be sufficient to offer to the experimental animals a favourable opportunity to compose and to retain shell water of a desired concentration.

After shaking out all shell water, normal animals from the storage box were placed in groups of ten, under conditions of 28°C and about 75% R.H., in the testing tank. Four or five days later the shell water was shaken out again and tested. A hundred and ten out of 120 tested animals had retained enough shell water to permit determination of the electrical conductivity.

A diagram showing the distribution of this shell water concentration, converted into salinities, is given in Fig. 20.

Although theoretically the salinity of the determined shell water

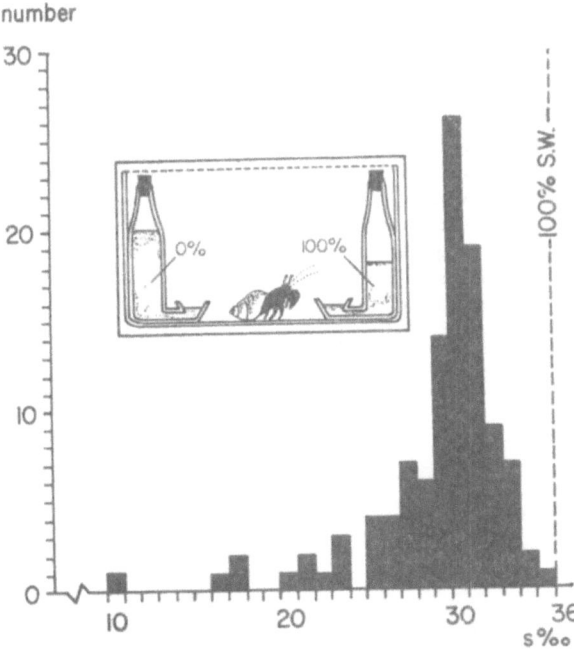

Fig. 20. SALINITY RANGE OF SHELL WATER in specimens of *Coenobita* which had access to distilled water (0% sea water) and normal (100%) sea water, under laboratory conditions. — Broken line: 100% S.W. *Inset*: experimental set-up.

concentrations might range from 0‰ S (dist. water) to over 36‰ S (normal sea water), the lowest value found here is a salinity of 10‰; the highest 35‰. A very distinct peak occurs at a salinity of 30–31‰ S. All salinities are less than 100% S.W.

Since a period of 4 or 5 days is much too short for building-up shell water concentrations as discussed in experiment 1 by the uptake and evaporation of low-concentrated water only, we have to conclude here again that in the first place the regulation consists of a simple mixing of water of different concentrations in the right proportions. A possible release or exchange of salts or ions by the animal is left out of consideration, though it very likely takes place.

98

Differences between small and large animals

Two additional experiments, fully identical to the one described in the last section, were performed with 2 groups of small- and of large-sized animals respectively (gross weights of the small animals ranging from 10–20 g; large ones ranging from 50–100 g). The intention was to investigate whether the size of *Coenobita* could be correlated with the concentration of shell water retained by the animals. Such a correlation was found indeed, as could be concluded from the results of two series of experiments in which shell water of a total of 120 specimens was tested. Small animals build up shell water of a somewhat higher concentration than larger animals do in the same period (Fig. 21).

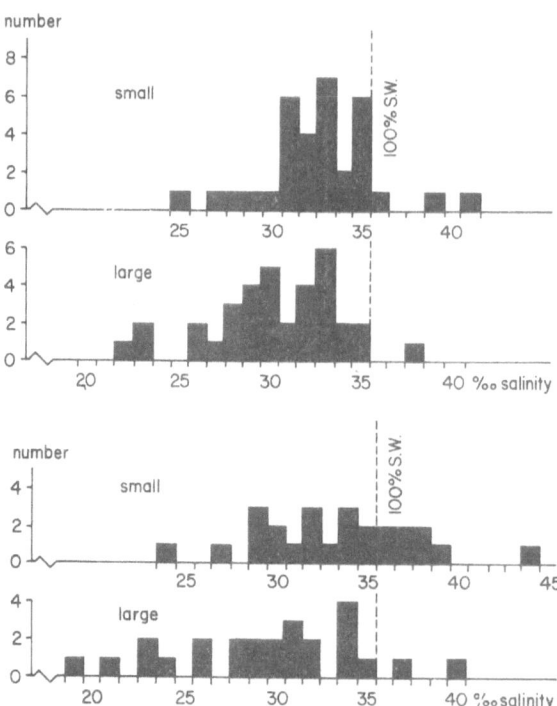

Fig. 21. DIFFERENCES IN DISTRIBUTION OF THE CONCENTRATION OF RETAINED SHELL WATER comparing small and large specimens of *Coenobita*. — Large individuals show distinctly lower values. Broken line: 100% sea water.

99

Mixing of other concentrations

As in the previous experiment, normal animals were allowed to choose between two concentrations 0 and 100% S.W., after all the shell water had been removed. In two more tanks combinations of 0% and 150% S.W., and 0% and 200% S.W. were offered to normal animals under the same conditions. After a period of 5 days the shell water of all animals was shaken out and measured. In addition an

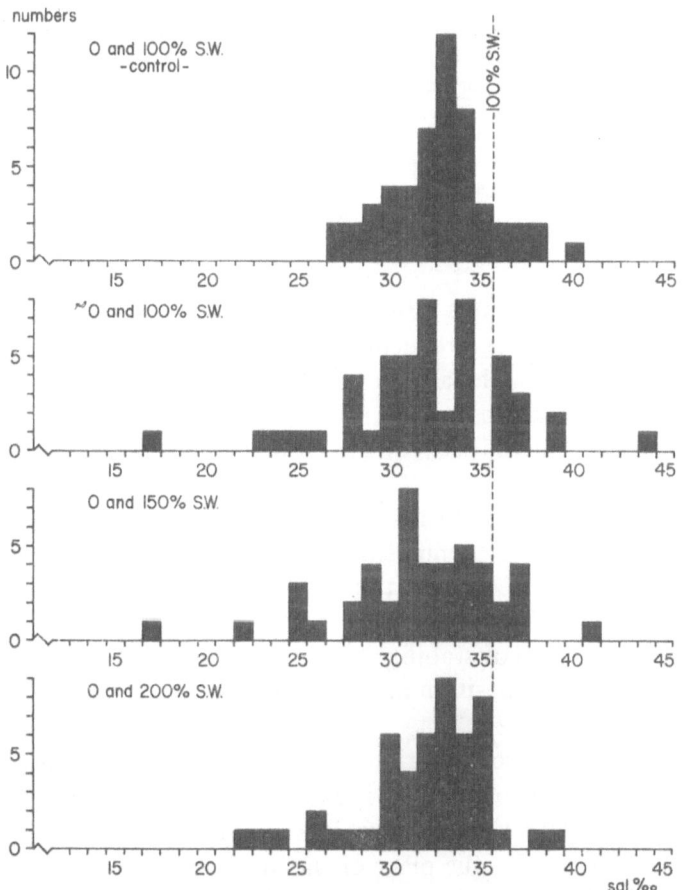

Fig. 22a. Equal distribution of shell water concentration in 8 groups of *Coenobita* having access to resp. combinations of 0 and 100, 0 and 150, and 0 and 200% S.W. — Control represents the shell water of animals from the storage box. — Measurements in first 4 groups after intervals of 5 days.

equal number of animals was kept back in the storage-box for control, where they also had access to both 0% and 100% S.W. The shell water of the latter animals was also examined at the end of the experiment.

As represented in Fig. 22a, the results in all cases show well-comparable frequency distributions of the salinity of shell water. In the same way as shell water is composed from 0% and 100% S.W., *Coenobita* is able to mix 0% and 150% S.W. or 0% and 200% S.W. in the right proportions to obtain the right kind of shell water. The most frequently occurring values for the salinity are those of about 32‰ S, which is distinctly below the salinity of 100% S.W. (S = 36‰), but somewhat higher than the values of 30–31‰ S, found for the shell water concentration in the previous experiments.

Moreover the distribution curves derived from the animals in the testing tanks show a somewhat larger spread and are more negatively skewed, as compared to the curve of shell water of the animals in the storage box.

The same experiment was repeated once more. The shell water of the animals in the three testing tanks, however, was now measured after intervals of one day. Moreover, shell water of an equal number of animals, kept in the storage box, was measured again for comparison. So the shell water of this last group, which had been kept in the box for at least a week was measured after a considerably longer interval.

Considering the results (Fig. 22b), the check (upper histogram) shows that the most frequently occurring shell water concentrations are those with salinities of 34–35‰, which is higher than the most frequently occurring salinities of the three lower histograms, which are only 31–32‰ S. Furthermore there is also a larger spread.

On the one hand the results confirm the previous findings, since an equally successful composition of shell water takes place if water of a different concentration in combination with fresh water is available; on the other hand, the factor time also seems to influence the degree of stability attained in the shell water. Over a longer time the composition and regulation of shell water seems to become more stabilized.

From the frequency distributions of shell water in *Coenobita* derived from the experiments, it may be seen that, although the figures show distinct maxima which are lower than the concentration of 100% S.W., the place of the maxima is somewhat variable, ranging from salinities of about 30–31‰ up to salinities of about 35‰ S. The same holds true for the total spread and consequently for the minimum and maximum values.

Assuming that there is only one concentration which represents

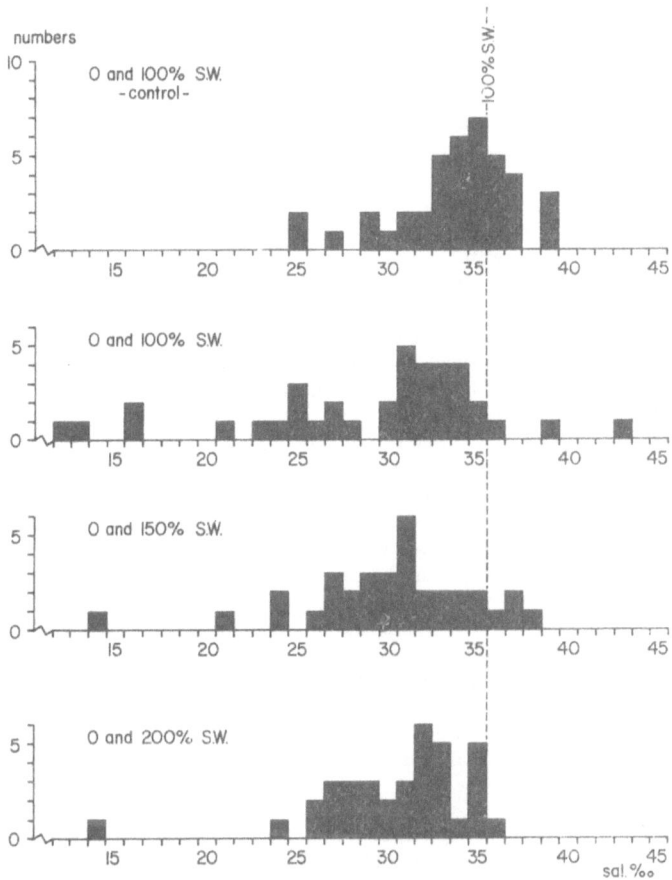

Fig. 22b. Same as Fig. 22a. — Measurements in second 4 groups after intervals of 1 day, except the control. Broken line: concentration of 100% S.W.

the optimal condition, the observed deviations in the found shell water concentrations, are caused by imperfections in the way of regulation itself, e.g. the attainment of the optimal concentration will take more time than the experiment permits. It is also assumed that the optimal concentration may change, depending on certain in- or external factors, such as the size of the animals, temperature, relative humidity, etc., or the combination of factors such as temperature and humidity.

Time as a factor influencing regulation

In a following experiment the influence of the time-factor on the final concentration of shell water was further investigated. For this purpose the concentration of shell water was determined for various groups of animals, staying under similar external conditions and having access to both 0% and 100% S.W. The time-intervals for the determination were 1, 2, 3, and 5 days. After determination of the concentration of shell water, the animals were discarded and replaced by new animals from the storage box, which were similar as to origin and previous adaptation. To provide an undisturbed uptake of water at the drinking fountains, never more than six animals were tested simultaneously in one tank. As a result of the restricted equipment, which did not permit the use of more than four experimental tanks, the animals in which the shell water could be measured after a 5-day interval, were not as numerous as those in which shell water was measured after shorter intervals. Thus in the first experiment, which lasted for 10 days, 58 animals were measured after 1 day, 29 after 2 days, 18 after 3 days and only 12 after an interval of 5 days. For a second series, lasting 15 days, the numbers in the groups were 74, 41, 30 and 17 respectively. In spite of this imperfection the average value of the measured shell water concentrations in each group increased from 30.3‰ S., over 30.6 to 32.0 and 31.9‰ S. in the first series and from 31.1‰ S. over 33.4 and 34.3 to 35.6‰ S. in the second series.

It is clear that shell water concentrations tend to become higher after longer intervals. In the first place this must be ascribed to evaporation. New water, however, is taken up by the animals in the course of time. As only relatively few animals were found with low shell water concentrations it must be assumed that the rise in concentration is also brought about by a more balanced uptake of water as the animals grew more familiar with the experimental tanks and the available shell water.

As the factor time has a (direct) effect on the shell water concentration, due to higher loss by evaporation, a similar effect may be expected to be caused by higher temperatures of the environment; lower relative humidity of the air and stronger air movements (wind) may as well affect evaporation positively. Uptake of the right amounts of drinking water of low concentrations will bring the concentration back to the correct value.

It will be clear that in this way a system of shell water regulation is encountered in which a fixed concentration may never be expected. Just after a dilution the shell water will have its lowest value; little by little, depending on the rate of evaporation, the concentration will rise. This will continue until the animals start to experience the concentration as being unfavourable. If possible they will now start to dilute the shell water again.

Somewhere between the lowest and highest values the shell water will have its optimal concentration. It is, however, not absolutely necessary that this optimal concentration coincides with the maxima in the frequency distributions of the concentration of the shell water.

Under favourable conditions most animals will be able to keep the differences between lowest and highest concentration of the shell water rather small. In Fig. 20, 88% of the measured values lie within the salinity range or 27–33‰ S. Preliminary experiments proved that the majority of the animals took up new water every night.

To get more information about the drinking frequency of *Coenobita clypeatus* the activities of over 200 tagged animals in the outdoor crab pen were followed continuously. Observations during part of the night in the end of April 1967, at rather dry weather conditions, indicated that as many as 50–65% of the animals visited the drinking trough between sunset and midnight and quite a few of them were seen twice near the trough. During the rest of the night a considerable number of animals would find their way to the water. Occasionally drinking animals were seen during daytime. From these data it was estimated that 75–100% of the animals took up water every day.

When circumstances get less favourable the differences between initial and final shell water concentrations will of course increase. For instance when the animal fails to dilute the shell water in time, the final concentration may increase to 45‰ S. As a rule, however, they will get rid of the water before this situation is reached. It is already known that water with salinities of over 36‰ (= 100% S.W.), was hardly ever retained as shell water; typical coastal animals were seldom found to carry shell water.

Apart from the question of the optimal concentration, which will be discussed later, other interesting data are the lowest concentration to which *Coenobita* does not object, and the highest concentration at which it is desirable for the animals to dilute or to remove the shell water.

As could be demonstrated in a subsequent series of experiments, the salinity of the initial concentration also depends on external environmental conditions. The ability to adapt the rate of tolerance, as far as the maximum (and minimum) concentration is concerned, must be very valuable indeed to the animals, as it will enable them to utilize a certain amount of water over a longer period under less favourable circumstances, but on the other hand it will make a great demand on their osmoregulation.

Influence of air humidity on regulation

This experiment describes the differences in concentration of the shell water in two groups of *Coenobita*, exposed to different degrees of evaporation, induced by high and low relative humidities of the air. Other factors, which are supposed to affect the evaporation, such as temperature (indoor temperatures of 28°C) and wind, were kept as constant as possible. As a suitable experimental set-up 2 similar tanks were chosen as described, but now both were enclosed by a much larger (80 × 50 × 50 cm) eternite container, covered with an air-tight plate-glass window. Through small holes in the side walls of the outer container cables were fitted for temperature- and humidity probes, which were mounted under the wire-mesh covers of the inner tank. Both types of probes were connected with a recorder. The bottom surface of one of the containers which was not occupied by the inner tank, was filled with a number of shallow trays with distilled water and provided with cottonwool evaporators reaching up to the roof of the container. On the bottom of the twin container shallow trays holding about 2 kg of dry silica gel were placed.

Under temperatures of 28 ± 2°C, the relative humidity in the wet tank fluctuated around $95 \pm 5\%$ R.H. after some hours; at the same temperature the R.H. in the dry tank was $40 \pm 10\%$. At night (24.00 h) six normal animals, with all shell water shaken out, were released in both tanks. Next morning the concentration of the shell water present, taken up during the night, was determined. Three nearly identical series of experiments, in which respectively 120, 48 and 72 animals were used, gave higher averages of the concentrations in the wet tanks than in the dry tanks. Results are given in Fig. 23. Apart from the differences demonstrated here in the measured concentration of shell water – which seem to be controlled by the external environment, in this case the difference in R.H. of the air – it was also interesting to notice the quite different behaviour of the two groups of animals. "Dry animals" proved to be much inclined to take up shell water. After each night the animals were found huddling in the corners of the inner tank. The shells were well-closed with chelipeds and ambulatory legs. During the removal of the shell water for conductivity measurements it was noted that the shells were filled to the brim with shell water. The "wet animals" on the other hand showed little inclination to retain shell water. Though the shells of these animals were moist to the touch on the outside, shell water itself was found to be present in considerably smaller quantities and often not at all. In the "wet animals" the number of unsuccessful shell water determinations due to too small quantities of shell water was 28 out of a total or 120 tested animals. In the "dry animals" this number amounted to only 8 out of 120 animals. In general the "wet animals" behaved much more fidgety; walking about or sitting with ambulatories widely extended.

As a more or less logical sequel to this experiment attempts were also made in a series of experiments to investigate the effect of various temperatures – 20°, 28° and 36°C, at constant relative humidities – on the composition and maintenance of the shell water concentrations.

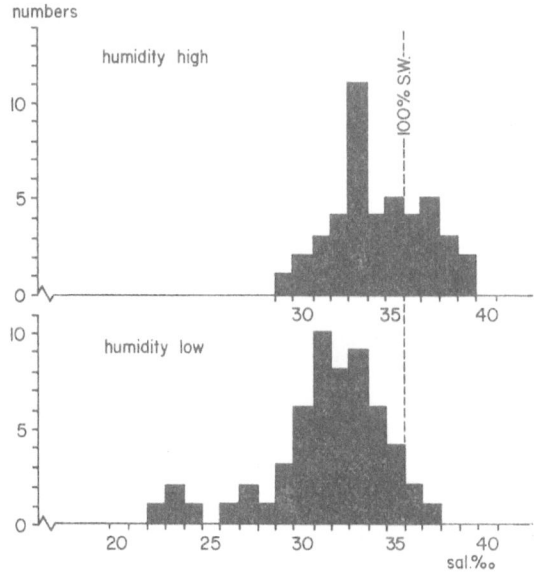

Fig. 23. DISTRIBUTION OF SALINITY OF THE SHELL WATER in *Coenobita* kept at high (95 ± 5%) and low (40 ± 10%), relative humidity both at 28°C. — Broken line: 100% S.W.

Due to the lack of suitable equipment in the Institute at Curaçao it was impossible to obtain a simple experimental device in which two or more different constant temperatures could be maintained under equal relative humidities. Several devices, utilizing evaporators of various sizes failed to give satisfactory results, for the small but unavoidable changes in temperatures always induced much larger changes in humidity.

As the effect of the R.H. on the concentration of the shell water had already been studied the experiments were not continued. Taking the observed lower concentrations in the previous experiment to be a reaction of the "dry animals" to the high evaporation caused by low humidity of the environment, higher temperatures are expected to cause a similar effect.

FIELD OBSERVATIONS

Compared with the laboratory experiments the number of animals in the field of which the concentration of shell water was measured is rather small. In 1966 on two occasions shell water was measured in small, medium and large animals near Punta Tera at Oostpunt, Curaçao. Most of them were collected in the gully which

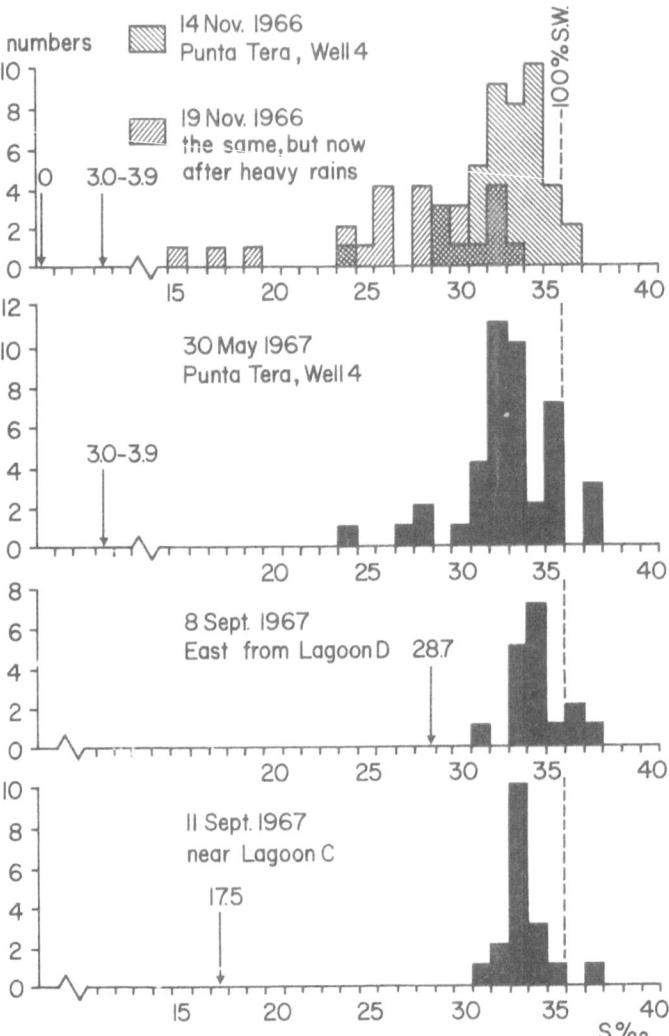

Fig. 24. DISTRIBUTION OF THE CONCENTRATION OF SHELL WATER from *Coenobita* collected in the field. — Concentration of shell water is regulated on a level of about 32–33% S. Heavy rains lower the concentration. Vertical arrows: concentration of available drinking water.

is situated eastwards of Well 4. In November the terrain was rather dry; 5 days later heavy rainfall occurred. The results of both sampling days are given in Fig. 24, upper histogram.

In 1967 shell water was collected in the field three times. On May 30 animals were collected near Punta Tera again, now in the plain immediately east of the well. Lots of animals fed here on the fruits of a kind of thron-apple (*Datura metel*). On September 8 shell water was sampled in a small number of Coenobitas eastwards of Lagoon D and on September 11 in animals congregating at the northern bank of Lagoon C. Histograms of the results of the last group of shell water measurements are also presented in Fig. 24.

Data on the shell water of *Coenobita clypeatus* sampled in Bimini, Bahamas, will be published separately.

As mentioned above salinity of the shell water of (inland-)animals sampled in some of the natural habitats in Curaçao ranges from 24‰ up to values of 37‰ S. The most frequently occurring values are those in the range of 31–34‰ S. Again the majority of the examined concentrations of shell water lies below the concentration of normal (100%) sea water in the Caribbean.

Due to heavy and frequent rain showers the shell water may show a decrease in concentration, as is obvious from the data of November 19.

FINAL CONCLUSIONS

The fact that terrestrial hermit crabs must be counted among the most successful land inhabiting decapods, is for the greater part due to the role played by the adopted shell. Besides some obvious advantages, such as protection against damage and predation, its chief functions are the prevention of desiccation and the retention of shell water. Under favourable circumstances the animal not only retains a considerable stock of water, but is also able to keep the salinity of that water at a more or less constant value of 32 to 33‰ S. The importance of water management for the animal is obvious. Moreover, in this way it creates an intermediate environment between internal and external media, which contributes greatly to the facilitation of osmoregulation.

It must be noted that by no means all hermit crabs under natural conditions possess shell water. As a rule small individuals, or animals living near the sea, do not store water. So it is not indispensable. On the other hand shell water is practically always found in larger individuals of the inland animals. Observations indicate that the variation in concentration of this shell water is sometimes limited (Fig. 24). The regulatory mechanism, however, often comprises much greater fluctuations. Besides, a more or less skew frequency distribution of the measured concentrations may be expected. Individually the concentration of shell water will show a gradual rise, followed by a sudden fall at every dilution, yielding a kind of serrated line.

From laboratory determinations on hundreds of animals it appeared that the lowest value was a salinity of about 10‰; the highest values were found at about 44‰ S. Various diagrams in Figs. 20, 22 a, b and 24 also show a distinctly skew distribution. Generally the diagrams show a fairly distinct upper boundary at 36–37‰ S. There is no clear lower limit. Depending on various causes, such as different sizes of the animals or differences in time available for stabilization, or in humidity of the environment, the maxima of the diagrams of the various experiments shift between about 30‰ and 35‰ S. It is by no means certain that those maxima also represent optimal values, values at which osmoregulation requires the least possible energy, or in which a constant gradient with the environment is maintained. The decisive factor is whether an animal can renew its shell water in time.

V. OSMOREGULATION

INTRODUCTION

For the greater part of their life the truly terrestrial crabs never
return to an aquatic life. Strictly speaking, these animals should
have no osmotic problem, since they lack an aqueous environment.
Sometimes, even on dry land, terrestrial crabs may still have a
concealed external environment in the shape of water retained in
the branchial cavities and, as is the case in hermit crabs, the shell
water. Continuous water losses, however, both from the animal
and the retained external storages, cause an increase in concentration of body fluids as well as of external water.

According to various authors terrestrial crabs are capable of
osmoregulation, or at least of maintaining the osmoconcentration
of the body fluids within certain limits. Depending on the environment in experiments or in the field they show either hypo- or
hyperregulation (see e.g. GROSS, 1957; BLISS, 1968).

Here the concept of osmoregulation is used in its widest sense:
that of keeping the body fluids at a fairly stable level, including
the possibility of hyporegulation in avoiding or diminishing high
concentrations, caused by evaporation. It appears that in adult
land crabs the gills, pericardial sac and gut are particularly involved
in the uptake, storage and redistribution of salt and water, whereas
the antennal gland are mainly concerned with the ion-regulation.

From all terrestrial species the terrestrial brachyuran decapods
have been most thoroughly investigated as far as osmoregulation
is concerned. Unfortunately far less attention has been paid to

terrestrial anomuran decapods and, though recent studies on other decapods may be of general application, the knowledge about terrestrial hermits is mainly based on older publications. Such references, however, are scarce.

In the Tortugas, PEARSE (1932) measured the depression in freezing point of the blood of *Coenobita clypeatus* and found that the range was 1.90 to 2.21°C (average 2.09°C), which is slightly higher than the value of 2.04°C for normal sea water with a salinity of 36.05‰ in that area. A species of the marine genus *Petrochirus* gave comparable values, from which the author concluded that both *Coenobita* and its marine ancestors have blood contents equalling normal sea water. For terrestrial brachyuran species, such as *Gecarcinus* and *Cardisoma*, however, distinctly lower values were obtained, ranging from 1.55 to 1.79°C.

In *Birgus* and *Coenobita*, GROSS (1955) measured various concentrations of body fluids, dependent on the salinity of the available drinking water. When both fresh water and sea water was offered to *Birgus*, the concentration of the body fluids ranged over a value equivalent to 85 to 92% sea water. When only fresh water was offered, concentrations equivalent to 65 to 74% S.W. were found; when only normal sea water was available, the concentrations in the body might increase to the equivalent of 118 to 128% sea water.

When given the choice between fresh water and sea water, animals with low blood concentrations prefer the latter, and with high concentrations the former. In this way *Birgus* will select the proper amounts of water of different salinities to adjust its body fluids till the optimal value for the species is reached. Terrestrial hermit crabs show a similar behaviour; as will appear from this paper, they are capable of adjusting the concentration of their retained shell water in a similar way.

Remarkably constant values for the blood serum of *Birgus latro* were found by HARMS (1932), who obtained freezing points of −1.42 to −1.49°C. In addition HARMS mentions comparable blood concentrations for two *Coenobita* species from the same area.

EXPERIMENTS AND OBSERVATIONS WITH COENOBITA CLYPEATUS

Attempts were made to study the concentrations of the blood, the urine, and also the shell water, as well as the osmoregulatory abilities of *Coenobita*. Moreover the relations between the osmo-concentrations and the environmental conditions as met by the animals in their natural habitats were investigated.

Fig. 25. RELATION BETWEEN WEIGHT LOSS BY DESICCATION AND BODY FLUID CONCENTRATION, Δ_1, as observed in four Coenobitas (a, b, c, d). — Solid lines: weight loss by evaporation; broken lines: increase in body fluid concentration derived from freezing point determinations (o) in °C; dotted lines: theoretical concentrations of body fluid in °C calculated from the concentration indicated by arrow and daily water loss; †: death of animal.

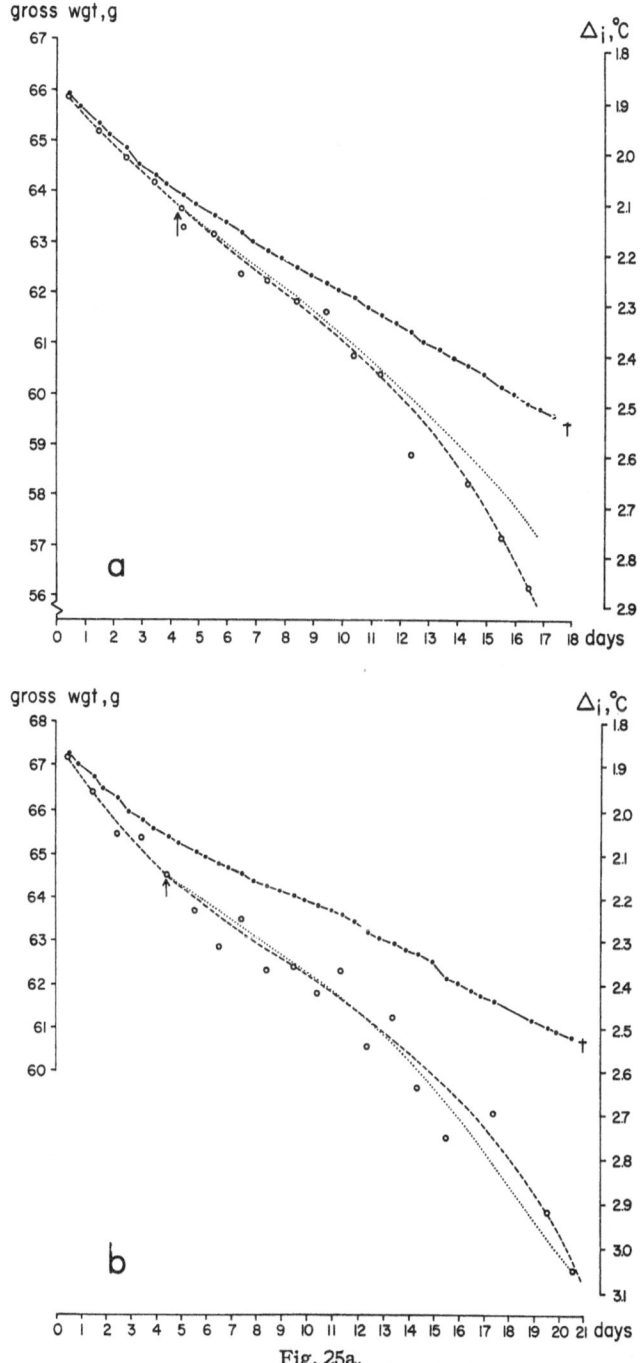

Fig. 25a.

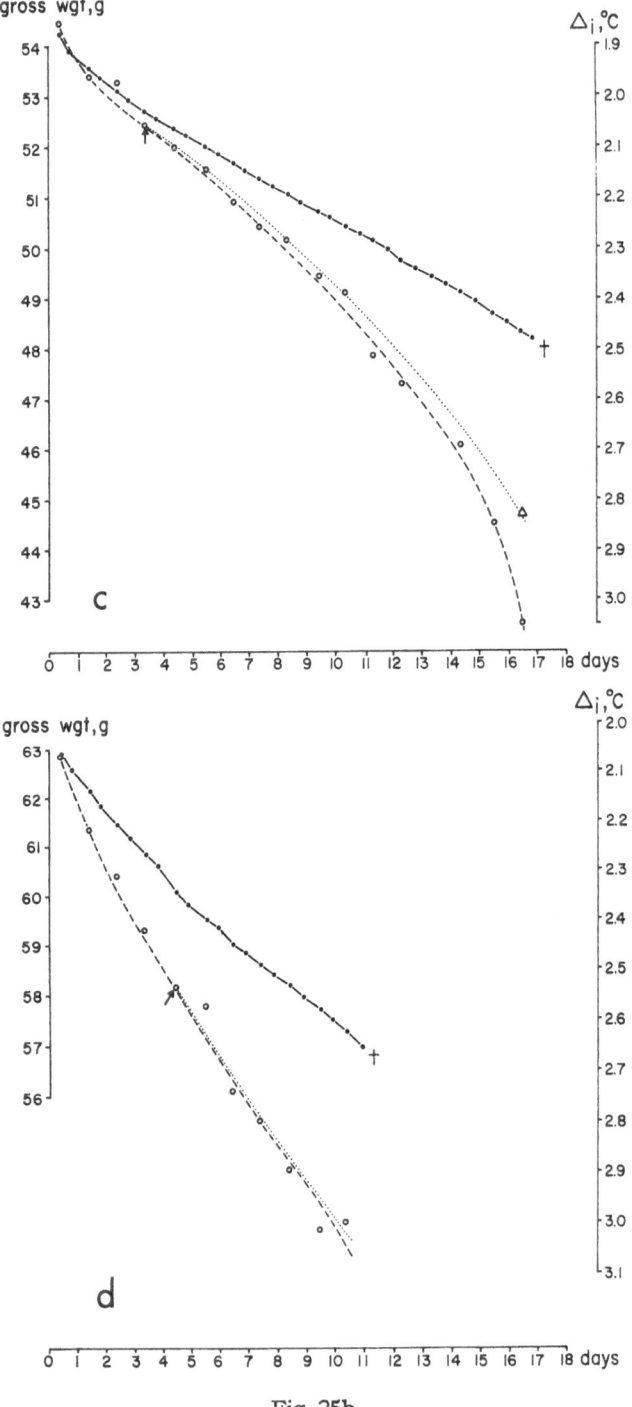

Fig. 25b.

Effect of dehydration on concentration of body fluid

In addition to the experiments described in Chapter III the effect of water loss by evaporation on the concentration of the body fluids of *Coenobita* was investigated in order to answer the important question whether these animals have some mechanism to lower the concentration of their body fluids.

It has already been demonstrated how, and at which rate, these losses occur and it was shown that a water loss equalling 15–20% of the net body weight is lethal. It is not directly clear, however, which organs are responsible for the hypo-osmoregulation, if no free water is present, neither in the branchial cavities nor in the shell.

Methods

In a number of desiccating animals the water loss under environmental conditions of 28°C and a R.H. of about 75% was found by determination of the decreasing gross animal weights twice a day. In addition a very small blood sample was collected every day for measuring freezing point depressions, by penetrating a membraneous joint of one of the ambulatories with a finely drawn out glass capillary. After sampling the animal was weighed again for determination of the weight loss caused by the sampling. As the blood samples averaged only 0.006 g each, this loss was left out of consideration in the further calculations. Each of the four experiments ended by the death of the animal. Then wet and dry weights of shell as well as the dry weight (24 h at 110°C) of the animal itself were determined. From this information the water content of the animal in each step of the desiccation process was calculated.

Results

In four cases Fig. 25 shows the decreasing gross animal weights as well as the increasing body fluid concentrations, Δ_1, as calculated from the freezing point depressions. Also the concentration curve as calculated from water loss is given. From the moment the water loss becomes more or less constant after the initial period of greater loss, probably due to loss of shell water, the theoretical values of the freezing points are calculated from the concentration (indicated by arrow) and the daily loss of water.

It is clear that the effect of (hypo-)regulation becomes visible when the measured curve deviates in an upward direction as compared to the theoretical curve, derived from water loss. Since, however, the two curves tend to coincide for the greater part, or even the measured curve lies below the theoretical curve, such a regulation is not apparent. Therefore a decrease in osmoconcentration, i.e. a reduction of the total number of particles in the blood, cannot well be attained by the animals under circumstances in which they are exposed to desiccation and free water is not available.

The effect of reviving

In the desiccation experiment discussed in Chapter III the increase in concentration of the body fluid as a result of dehydration was determined by conductivity measurements. After ten days of desiccation, the animals had lost approximately 15% of the net body weight. The body fluid concentrations, as far as the electrolytes are concerned, had increased up to the remarkably high values of 170% of the initial concentration. Then fresh water was offered, which was taken up avidly. Determination of the net body weights showed that the animals had regained or even surpassed their original weights within two days. The body fluid concentration had again reached the original values (Fig. 26).

DISCUSSION

From these findings, combined with those of the previous experiment, it may be concluded that osmoregulation in *Coenobita clypeatus* stands or falls with the presence of water. It seems as if the animals are only able to control the concentration of the body fluids when free water is available, either taken up directly or retained as shell water. Whether this happens by a direct uptake by the mouth parts and penetration of the walls of the alimentary tract, or by an exchange of water and salts from the body fluids against water or salts in externally stored water, is still uncertain.

Thus is the absence of water, the animals must at any rate be

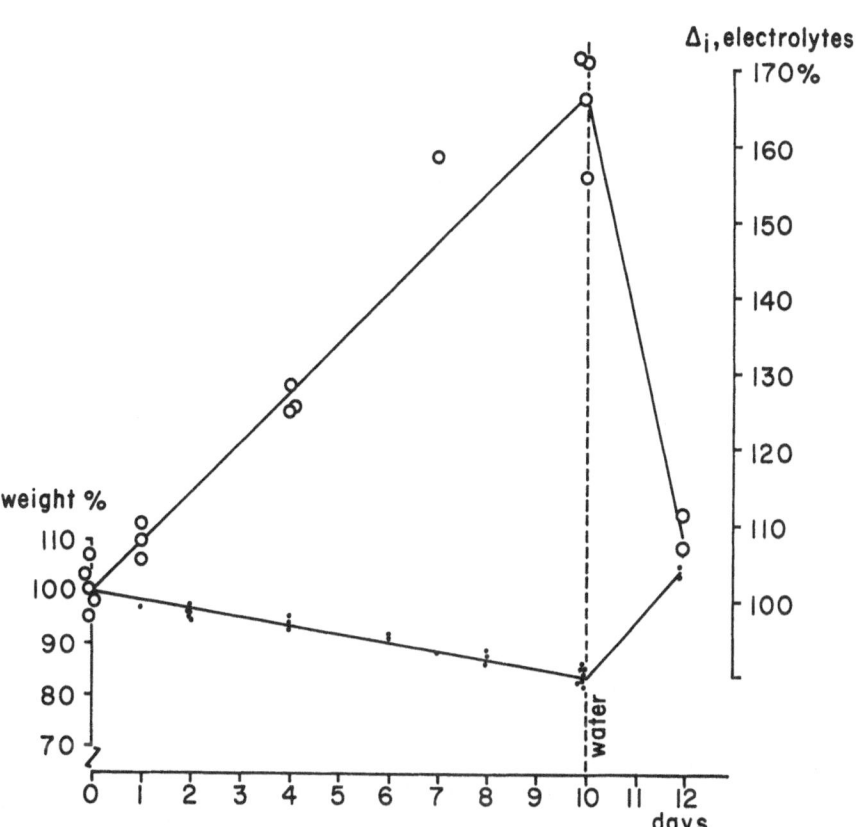

Fig. 26. CHANGES IN BODY FLUID CONCENTRATION, Δ_i, as an effect of water loss and reviving. — Net body weights (dots) and electrolyte concentrations (circles) derived from conductivity measurements expressed in percentages of initial values (= 100%). After 10 days of desiccation water was offered.

able to tolerate rather wide variations in internal concentration or in other words behave more or less like osmoconformers.

When water is available the animals will turn to osmoregulation and behave like osmoregulators. From the observations by GROSS (1955) on *Birgus* it is clear that, even with water of different salinities available, considerable variations in body fluid concentrations may still occur.

Osmoregulation, when water in any form is present

As osmoregulation appears to fail or at least to be of minor importance in animals deprived of water, the osmoregulatory capacities in terrestrial hermits having access to water will be discussed first.

The existing literature (GROSS, 1955) on the capacities of the related species *Birgus* records rather diverging concentrations, depending on the concentration of the available drinking water.

A good deal of useful supplementary information on this subject was already obtained during my stay in Curaçao in 1961. Unfortunately in that period neither a cryoscopical device, nor a micro-method for conductivity measurements were available. Therefore only determinations which required at least 0.5 ml blood could be made, which meant that continued series of measurements with the same individual were practically out of the question. In general, experiments tended to require a good number of animals, since every determination caused the death of the animal.

A preliminary experiment already demonstrated that GROSS' observations also held for *Coenobita clypeatus*. When drinking water of various salinities is offered to – somewhat desiccated – hermits, a notable change in the internal fluid concentration results immediately. When normal seawater is offered the concentration of the body fluids increases with 4–20% within a lapse of one day. When on the other hand distilled – or tap water is offered, this will in the same time induce decreases ranging from 4 to 24% of the initial values. From the rather quick effect of the absorbed drinking water the impression is gained that water and salts will pass the walls of the alimentary tract very quickly.

Of course experiments of such a short duration will only give very limited information. In the first place the amounts of water absorbed differ largely, depending on the salinity of the water. Moreover a certain amount of shell water may be retained and when it is not, the osmotic possibilities of the abdominal tegument are for the greater part ruled out, if indeed they play any role at all. To avoid these complications, at least to a certain extent, the animals were placed in tanks, filled with a layer of about 2 cm of water of a given concentration, so that the shells became filled and the entire body of the crab was sufficiently bathed in the water. On the other hand enough of the animal's body emerged to warrant a normal function of the respiratory surfaces. From their behaviour it was clear that the animals were not comfortable. Most of them tried to extend their ambulatories well over the surface of the water, or they would try to climb on top of each other to get out of the water. From time to time clusters of animals had to be disentangled.

The effect of osmotic stress on mortality

To study the effect of various osmotic stresses on *Coenobita* a mortality test was carried out. A number of tanks were filled with water of a salinity of 35, 40, 45, 50, 55, 60, 65, 70, 80, 90, 100, 110, 120, 130, 140, and 150‰ resp., and ten animals were placed in each. Every day the salinity of the water was checked and adjusted if necessary. At first the number of dead animals was counted at intervals of 30 minutes, later on at longer intervals. Animals were considered to have died when they did not react by pinching, when the hairs inside the large nipper were touched, and moreover, when they could easily be removed from the shells. The experiment was continued for 45 hours. The temperature was kept at 28°C.

The main result was that all animals in concentrations of 35 and 40‰ survived. In 50‰ the survival rate was 50%, in concentrations of 70‰ and over, all the animals died. In the highest concentrations the first animals were found dead after $1\frac{1}{2}$ to 2 hours. After about 8 hours the general picture of mortality was practically stabilized.

At concentrations of 90‰ and more the mortality was 100%, the animals dying more or less simultaneously after an interval of 6–10 hours (Fig. 27).

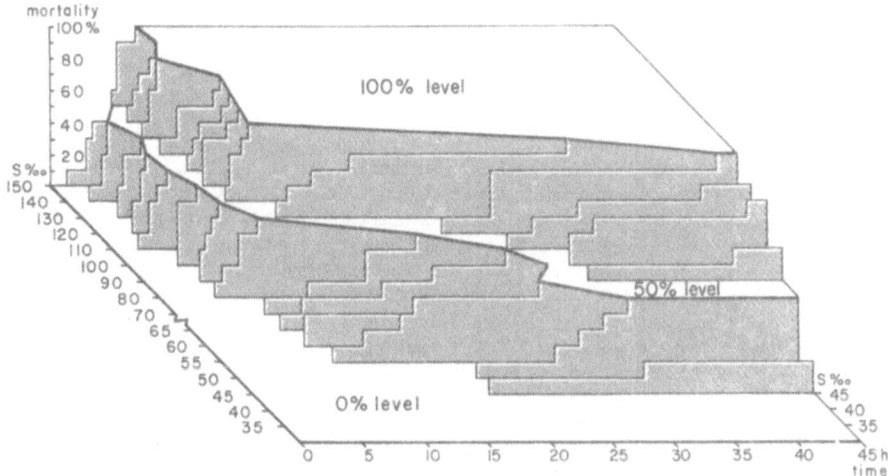

Fig. 27. MORTALITY IN *Coenobita clypeatus* AT VARIOUS SALINITIES of an imposed external medium as a function of time.

From these results it is clear that against osmotic death, supposed to be caused by permeating salts or by water loss, the tegument of *Coenobita* gives only a very limited protection.

In this connection it is interesting to stress that in similar experiments with diluted seawater mortality only occurred after much longer periods. Even a two-days stay in repeatedly renewed distilled water did not result in any significant mortality. (Compare also the treatment of the so-called "aqua dest"-animals). Since the cause of death becomes controversial, the long-lasting experiments in diluted sea-water are not discussed here. Yet they might lead us to suppose that either tolerance against dilution of the body fluids is much greater or that the prevention of either withdrawal of salts or penetration of water in the reverse direction is more effective than in the opposite case.

Still the concentrations of the body fluids are seriously affected when the animals remain under imposed conditions of osmotical stress.

The next series of experiments will show how in this case the osmotic concentrations of the body fluids change in the course of time.

The effect of osmotic stress on body fluid concentration

Use was made of external concentrations equivalent to salinities of $0.1\%_0$ (distilled water), $\pm 1\%_0$ (tap water), $36.0\%_0$ (normal sea water) and $50\%_0$ (hypersaline sea water). In the 175 animals originating from Klein St. Joris, Curaçao, February 1961, and used in the above-mentioned concentrations of 0.1, 1.0, and $36.0\%_0$, the initial body fluid concentration, as found from conductivity measurements, averaged an equivalent salinity of $22.5\%_0$. In another fifty animals from the same locality, but collected two months later and used in hypersaline sea water of $50\%_0$, the initial concentration averaged a salinity of $25.5\%_0$. The experiments lasted 7 days. Every day the water in which the animals were kept was removed and renewed, and the concentrations of both the old and the fresh water determined. After periods of 1, 2, 3, 4, and 7 days respectively, ten animals were removed at random from each of the four aquaria and the conductivity of their body fluid was measured.

The results of these experiments are summarized and plotted in Fig. 28.

It appears from the diagrams that the osmoconcentration of the blood, as far as the electrolytes are concerned, undergoes considerable changes as a result of the different environments. Distilled water and tap water induce a decrease in concentration, normal

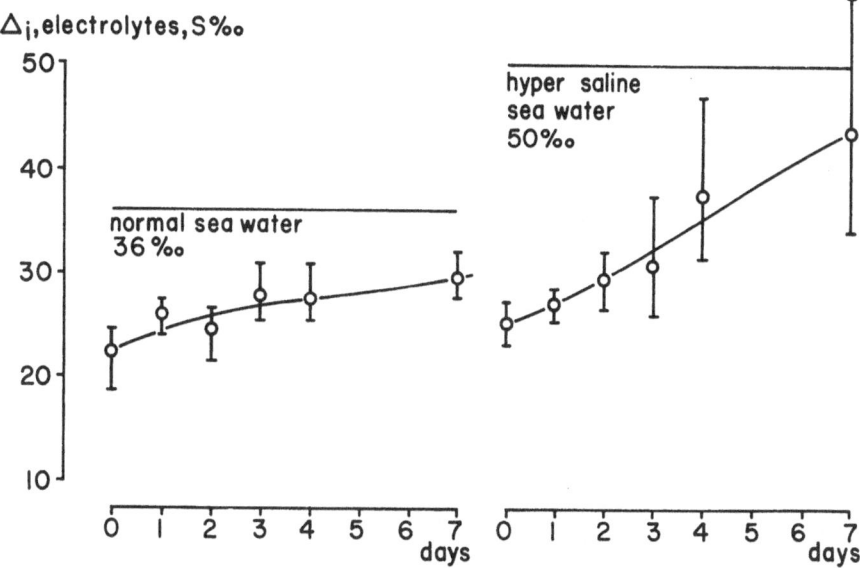

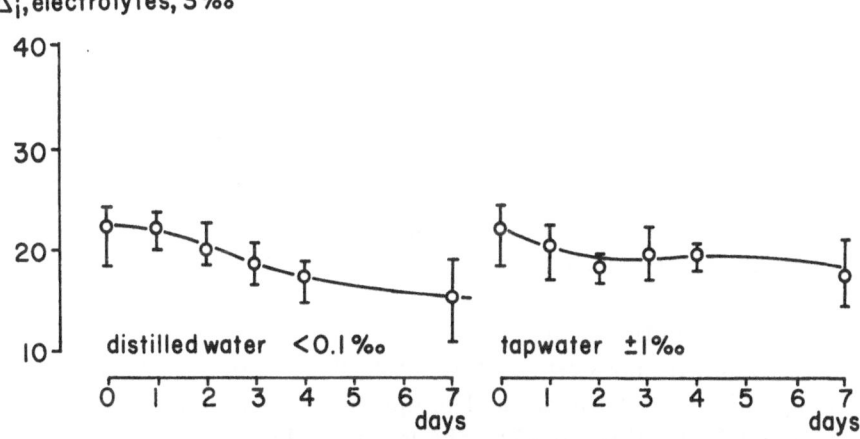

Fig. 28. OSMOTIC CONCENTRATION OF BODY FLUID, Δ_1, electrolytes of *Coenobita*, expressed in equivalent salinity in ‰, when various osmotical stresses are imposed — Indicated are the mean, and the ranges on either side of the mean.

sea water and hypersaline sea water an increase in concentration. After an initial fall or rise, respectively, the lines for tap water and normal sea water tend to run horizontally, pointing to osmoregulatory abilities. A tendency to stabilize the blood concentration at a new level is far less apparent with the animals in distilled water and absent in hypersaline sea water.

For animals in sea water the new level varies around an equivalent salinity of 30‰, for animals in tap water around 17‰. It is to be doubted whether such an equilibrium will be found with animals in distilled water. On the seventh day of the experiment there is still a considerable spread in the values in the diagram, and moreover a heavy mortality begins to appear. With animals in sea water and tap water no mortality occurred. With animals in hypersaline sea water the curve starts to rise quickly right from the start, and continues to rise during the next days. After the fourth day, when the average concentration of the electrolytes in the blood is about 40–45‰ the mortality is very great indeed. On the seventh day, the majority of 15 animals which had been kept in reserve for replacement, and which were kept in the same environment, had died, and it was very difficult to measure the conductivity of the blood of the remaining animals. The abdomens of these crabs had shrunk considerably and held only very small quantities of blood.

Two important conclusions may be drawn from this experiment: in the first place that the animals possess abilities for both hypo- and hyper-osmoregulation, provided that they stay in an aqueous environment, and secondly that a considerable range in the concentrations of body salts can be tolerated, as concluded from conductivity measurements, and therefore only concerning the electrolytes.

High mortalities in the distilled water and hypersaline sea water series indicate, however, that there exists an upper as well as a lower limit to this tolerance. The upper limit is found at an equivalent salinity of about 45‰, the lower at about 15‰ S.

As has been stressed it must always be kept in mind that the above data were found under conditions of osmotical stresses imposed upon the animals. In its natural surroundings *Coenobita clypeatus* will never be faced with such extreme conditions. The animals are never to be found in water, and shell water of either low and high concentrations is avoided.

This is why another series of so-called endurance tests was carried out, in which the long term effects were studied of a situation in which the crab was offered drinking water of a fixed salinity over long periods. It was left to the animals whether they would imbibe the water offered, retain it as shell water, or leave it alone. Precautions were taken to prevent the crabs from falling into their drinking bowls and thus creating a situation as had been tested in the previous experiment.

Endurance test 1

In the first series three types of drinking water were offered, viz. fresh water (S about 1.0‰), normal sea water (S 36.0‰) and hypersaline sea water (S 50.0‰). In large glass tanks the bowls, covered with a grid, were placed in such a way that the crabs could only touch the water with their chelae. In the freshwater test 5 animals were tested over 105 days, the other two tanks contained 10 animals each. The animals were tested in normal sea water for over 105 days, those with hypersaline water for 43 days. Every two days the gross weight of each crab was determined, to be converted into the net weight afterwards and then again into the percentage of the initial net weight. Fresh water and sea water were renewed every day, hypersaline water every two days. The temperature was kept at 28°C and the relative humidity at about 75%.

The results of the experiments are represented in figures 29, 30 and 31, summarizing the weight percentages and including several other data, such as shell exchange, mortality, cannibalism, etc.

On the first day practically all the animals show a definite increase in weight, as water is taken up. The fluctuating course of the graphs may be explained as follows: 1. Water uptake is followed by loss through evaporation (viz., in the parts of the curves where the weight decreases gradually for some days). 2. It is known that considerable quantities of water may be stored as shell water. It is hardly avoidable that some of this water is lost when the animals are being weighed. The greater part of the sudden drops are due to this loss. 3. When two animals exchange shells mutually, the altered relation between volume of animal and volume of shell may often result in a marked change in weight. In the figures such changes are indicated by a vertical dash. 4. Other causes of different nature, e.g. self-amputation of legs, are also indicated.

In Figure 29 (fresh water) the weight curves show an obvious descent in the course of time. This is especially clear with the animals 1, 2, 3 and 5. The drawn lines have been drawn by sight. During a period of 105 days 4 out of the 5 animals died and three exchanges of shell took place.

In Fig. 30 (normal sea water) several different weight curves show a horizontal

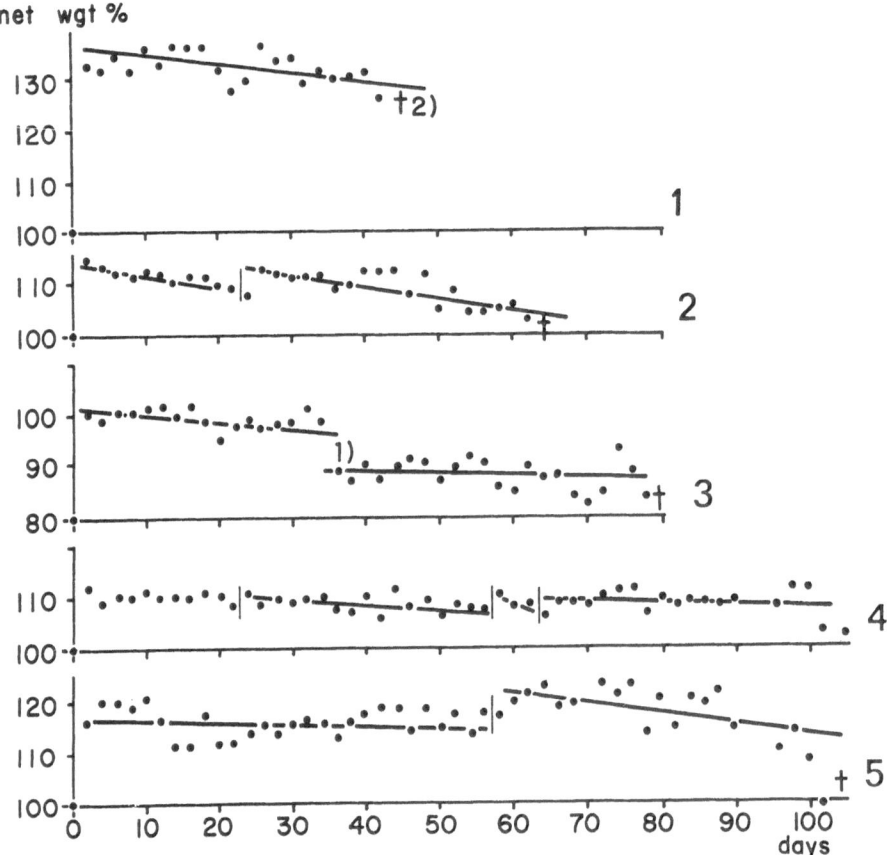

Fig. 29. ENDURANCE TEST WITH FRESH WATER (S = ± 1.0‰). Net weight in percentages of 5 Coenobitas versus time. — After initial increase of weight by drinking during the first days all individuals gradually show loss of weight. Vertical dashes: shell exchange; †: death of animal; 1): self-amputation of ambulatory; 2) death by cannibalism.

Fig. 30. ENDURANCE TEST WITH NORMAL SEA WATER (S = 36.0‰). Net weight in percentages of 10 Coenobitas versus time. — Various weight curves show horizontal or slightly upward trends. Vertical dashes: shell exchange; †: death of animal; 1): self-amputation of ambulatory.

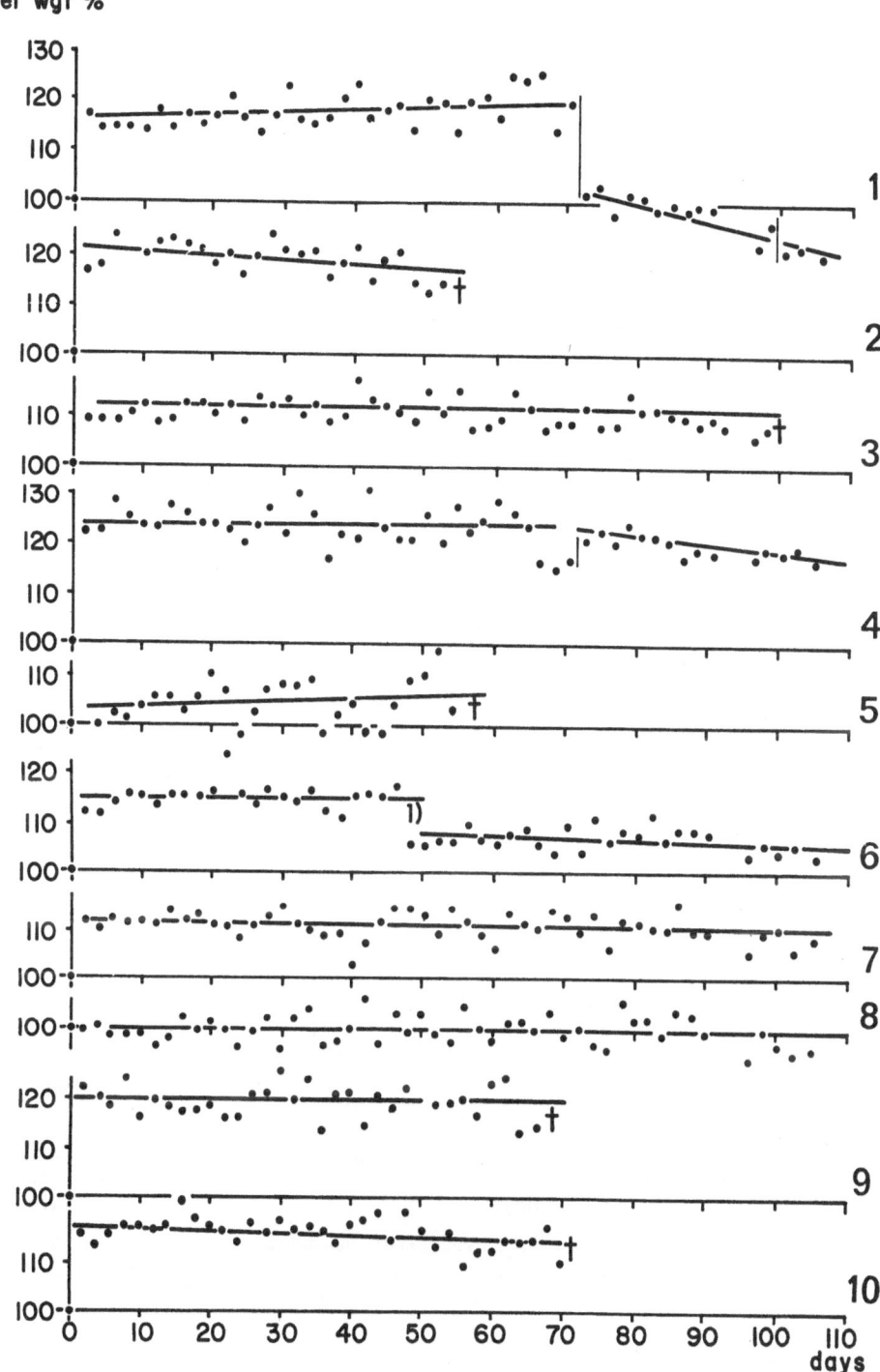

124

or even a slightly upward trend (drawn line). In 105 days 5 out of the 10 animals died, 2 exchanges took place.

In Figure 31 (hypersaline sea water) practically all the curves show a downward trend. Five out of the 10 animals died within a period of only 43 days. Three exchanges of shell took place.

A similar test in which six animals were offered hypersaline sea water of a salinity of 120‰ resulted in the death of all animals within 24 hours.

When the results of these endurance tests are compared the conclusion may be drawn that under the given circumstances normal sea water offers the best chances of survival of these ani-

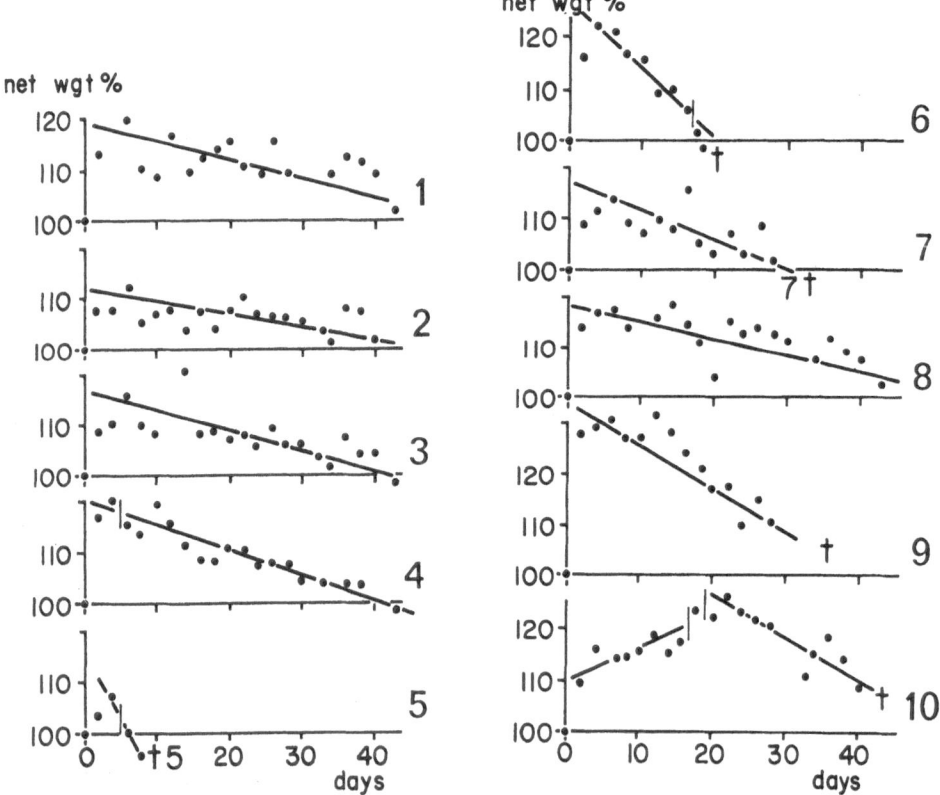

Fig. 31. ENDURANCE TEST WITH HYPERSALINE SEA WATER (S = 50‰). Net weight in percentages of 10 Coenobitas versus time. — Nearly all individuals show a rapid weight loss. Vertical dashes: shell exchange; †: death of animal.

mals. Loss in weight and a tendency to exchange shells are also assumed to be symptoms of a lowered well-being.

Endurance test 2

This test was set up with a greater diversity in the drinking water offered, since water of concentrations of 0, 10, 50, 100, 150 and 200% S.W. was offered and moreover all series were set up in duplicate, one series at 22°C and one at 30°C. It was tried to keep the humidity as constant as possible during the experiment, but under the given laboratory conditions in could not be avoided that the extreme climatic conditions caused the humidity to fluctuate between 70 and 85% R.H.

Here all weight alterations agree with those of the previous experiment. Both 0% S.W. and hypersaline water cause considerable loss in weight. When water of 10, 50 or 100% S.W. was available, the animals were able to maintain a more or less constant weight over a prolonged period. Differences between the two temperature series are not decisive.

The survival times from Table 13 give more information. Again hypersaline sea water proves to be unsuitable for survival; 200% S.W. is worse than 150%. On the other hand 0% S.W. yields remarkably long survival values. The influence of temperature is of particular interest; there are obvious differences between the two series. In the colder series, with low concentrations of sea water (10 and 50% S.W.) survival times are twice as long as in the warmer series. With 100% sea water survival is about the same for both temperatures. With hypersaline sea water of

TABLE 13

ENDURANCE TEST OF COENOBITA CLYPEATUS AT TWO TEMPERATURES

Available water in concentrations of 0, 10, 50, 100, 150, and 200% S.W. Concentration of body fluid, Δ_1, expressed in equivalent salinity in ‰ by conductivity and by freezing point depression (av. of two determinations).

	mean survival time in days		general trend in weight change		Δ_1 by cond.		Δ_1 by F.P.D.	
	22°C	30°C	22°C	30°C	22°C	30°C	22°C	30°C
0% S.W.	248	153	↘	↘	20.0	23.8	29.5	29.3
10% S.W.	253	136	→	→	25.1	29.5	32.1	33.5
50% S.W.	326	164	→	→	28.1	30.9	34.4	35.6
100% S.W.	174	173	→	→	32.7	34.2	44.6	40.4
150% S.W.	35	69	↘	↘	—	—	—	—
200% S.W.	18	16	↘	↘	—	—	—	—

150% the survival time is twice as high in the higher temperature. It is not surprising that this trend is not continued in the last and most concentrated solution, since this high concentration is wholly unsuitable for drinking water, as is proved by the short survival period for both temperatures. So, with certain reserves, we may conclude that low concentrations of sea water are better endured in combination with (relatively) low temperatures and that high (but not too high) concentrations are better tolerated if the temperature of the environment is high. These results are opposite to the findings in *Crangon crangon* (L.) (e.g. SPAARGAREN 1971) where combinations of high temperatures and low salinity, and low temperature and high salinities are better endured.

When considering the data of blood concentration Δ_i, as derived from freezing points and conductivity and both expressed in equivalent salinities, in the first place an increase in concentration with increasing salinity of the available water is obvious. Moreover regulation is evident from the horizontal parts of the curves

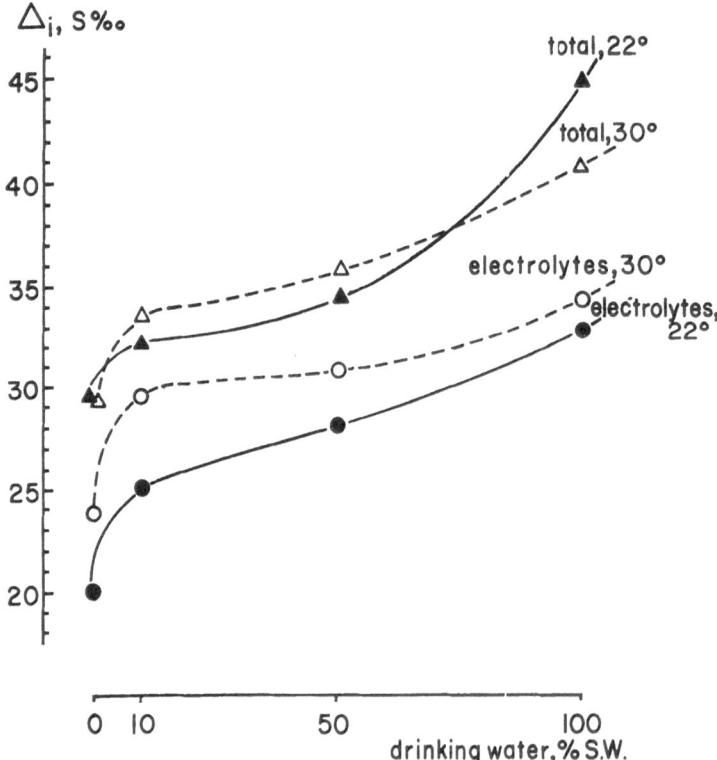

Fig. 32. TOTAL OSMOTIC CONCENTRATION AND ELECTROLYTE CONCENTRATION OF BODY FLUID in *Coenobita* expressed in equivalent salinity in ‰, at two temperatures as a function of the concentration of the available drinking water. — Concentration of drinking water expressed in percentages of sea water. After endurance test 2.

(Fig. 32). At higher temperatures the position of the two curves representing electrolyte concentration and total particle concentration is higher than of the same curves at the lower temperature. At 100% S.W. only the curve for total concentration shows the reverse.

The non-electrolyte component of the blood

The principal electrolytes in the body fluid are the inorganic ions Na^+, K^+, Ca^{++}, Mg^{++}, Cl^- and SO_4^{--} (ROBERTSON, 1960; PROSSER & BROWN, 1962). The total amount of electrically active particles, including the small amount of electrically charged organic molecules, can be measured by determination of electrical conductivity. Though important, the electrolytes contribute to only part of the total osmotic concentration, since this is connected with the total number of dissolved particles.

The non-electrolyte component, formed by proteins, free amino acids, (lipids and carbohydrates) cannot be neglected. Since the freezing point depression also depends on the total number of dissolved particles, irrespective of their nature, these data are most useful in this respect. By comparing freezing point depressions and electrical conductivity, the contribution of non-electrolytes to the osmoconcentration of the blood can be easily estimated.

Fig. 33 gives the concentration of body fluid Δ_i, as calculated from conductivity measurements and freezing-point depressions, for animals in which the concentration of their shell water is known. This shell water is considered to form the actual external environment, Δ_e of *Coenobita*. The curves in this figure have been drawn at sight. The distance between the curves, as measured along the vertical, represents the share of the non-electrolytes. Thus it may be concluded that non-electrolytes account for about 30% of the osmotic concentration; a percentage which is remarkably high as compared to data obtained with other crustaceans.

It is tempting to speculate on the conformity with the most successful group of terrestrial arthropods, the insects, in which comparably high fractions of organic substances, particularly amino-acids, contributing to the total osmoconcentration, are found (PROSSER & BROWN, 1962).

GRIMM (1969) states that proteins and other non-diffusable solutes in the blood plasma of 2 species of shrimps are probably present in low concentrations only.

128

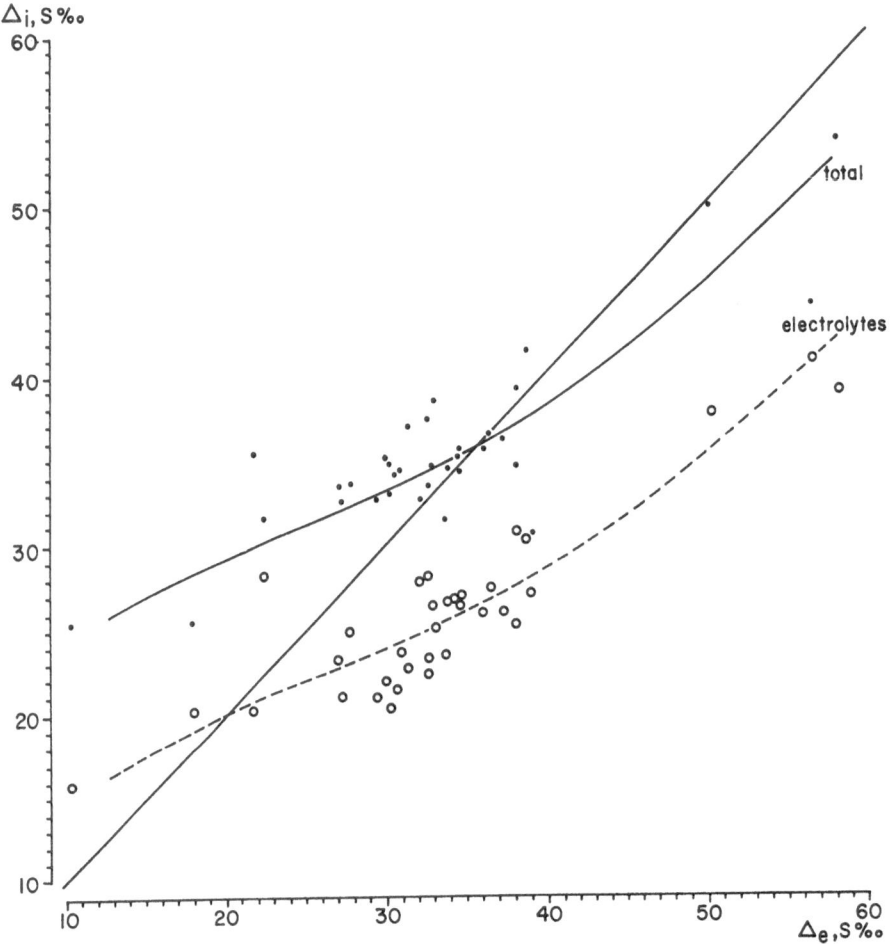

Fig. 33. BODY FLUID CONCENTRATION, Δ_i, total and electrolytes, both expressed in equivalent salinity in ‰, as a function of the concentration of the retained shell water, Δ_e, also expressed in S‰.

PARRY (1954) estimated protein in the blood of *Palaemon serratus* to amount to about 12%, a high value when compared with other decapods (ROBERTSON, 1960; vide GRIMM, 1969). According to SPAARGAREN (1971) the concentration of non-electrolytes in the blood of *Crangon crangon* and *Crangon allmanni*, depending on temperature and salinity, shows variations of 0–20% of the total osmotic concentration.

Osmotic concentration in the blood

It is also apparent from the deviation of the curves from the isosmotic line in Fig. 33, that apart from the regulation by means of the shell water itself, the abilities of regulation of the body fluids (Δ_i) against the shell water (Δ_e) also play an important part.

From these figures evaluation of the exact shape of the regulation curves and the effect of temperature on osmotic regulation is not well possible, notwithstanding the fact that data from animals kept at temperatures of 22, 28 and 38°C are included. In the area with low concentrations of shell water the osmoconcentration of the blood is hypertonic, in the area of high concentrations of shell water the blood is hypotonic. Since, however, the concentration of the stored water is also subject to regulation and the majority of the animals are able to maintain a shell water concentration of about 32–33‰ S, it may be said that in most animals the osmo-concentration is slightly hypertonic to the external environment (shell water) and hypo- or more or less isotonic to normal sea water. Furthermore, because the curves in Fig. 32 run parallel, it appears that osmoregulation is practically completely due to the regulation of the electrolytes, while the absolute concentration of the non-electrolytes hardly changes.

Isotonicity of the urine

From experiments in which the concentration of the body fluids as well as of the urine of desiccating *Coenobita* were measured by means of freezing point determinations, it was proved that the antennary glands of these animals do not have any osmoregulatory function. Body fluids concentrations expressed in equivalent salinities increase as result of water loss by evaporation; while at the same time determinations of urine concentration showed exactly the same rise. Figure 34 demonstrates the isotonicity of the two liquids.

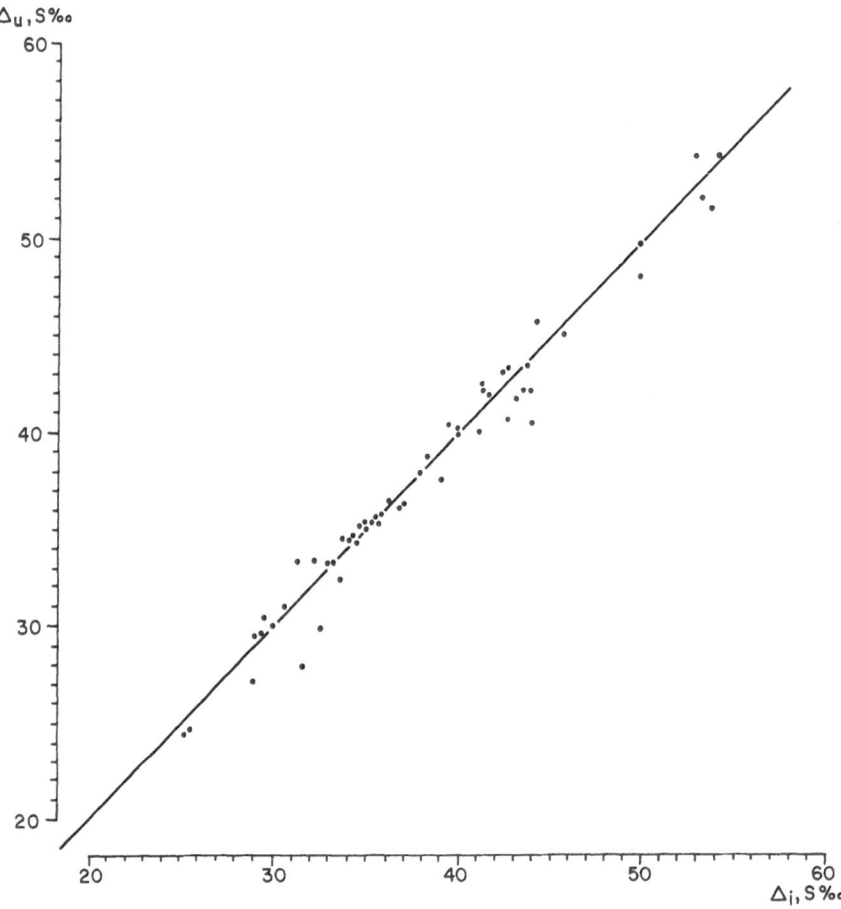

Fig. 34. CONCENTRATION OF URINE, Δ_u, of *Coenobita* at various fluid concentrations.

FINAL CONCLUSIONS

Terrestrial hermit crabs exhibit a remarkable variation in osmoregulatory possibilities. Dependent on environmental conditions they may be fairly good regulators, trying to keep the concentration of body fluids as constant as possible, both in hypotonic and hypertonic media. When, however, a moist external environment is not available, the animals are conformers, able to withstand enormous variations in internal concentrations.

Here osmoregulation is superimposed on regulation of the shell water, which is the only aquatic environment with which the animal is concerned, once its larval development is completed. In view of the variations in salinity of this shell water – from about 10 to 44‰ – and a general average blood concentration which for terrestrial crabs is slightly hypotonic to normal sea water, both hypo- and hyperregulation have adaptational values for this group of animals. A well-functioning regulation of the shell water will restrict its concentrations to values from about 30 to 35‰. Once this is obtained it is easy to maintain an equivalent body fluid concentration of 34 to 35‰ S by osmoregulation.

If no shell water is present *Coenobita* has no adequate means of hypo regulation. Progressive desiccation, which may go as far as the loss of about 30% of the internal water at the utmost, causes the concentration in the body fluids to rise to incredibly high values, equivalent to a salinity of over 60‰. A great number of adaptations to restrict water loss as well as a wide range of possibilities to make use of even the lowest amount of suitable drinking water protect the animal against such exceptional concentrations.

SUMMARY

r. This paper deals with various aspects of the life-history, ecology, water management and osmoregulation of the West-Indian land hermit crab *Coenobita clypeatus* (Herbst) in Curaçao, Netherlands Antilles.

2. Land hermit crabs belonging to the family Coenobitidae may be considered as one of the most terrestrial forms of decapods. They are characteristic for tropical coasts and islands. Though *C. clypeatus* may be found in a variety of habitats they show a preference for areas with a relatively dry climate. In this respect habitats as found on the Leeward Group of the Lesser Antilles are representative for the species' occurrence. In addition to populations of animals living on the coast, the 'coastal animals,' there are also individuals living in the interior, the 'inland animals.' The latter generally are older specimens, living in well-fitting and undamaged *Livona*-shells, and able to settle and maintain themselves in habitats rich in food, where there is a supply of fresh or brackish water. Though these animals still maintain a close bond with the sea, they may be said to have reached a more advanced stage of terrestrial life. The greater part of the research was carried out with inland animals.

3. In July reproductive migration starts, adult and not yet fully grown animals migrating towards a restricted number of spawning places on the southern coast, probably following traditional pathways. These places are characterized by the presence of good shelter, suitable drinking water and a quiet and accessible coast. Usually animals of roughly the same size gather in separate groups. It is assumed that fertilization occurs here. Soon afterwards, usually around the time of full moon, the first ovigerous females may be observed. Under the circumstances prevailing on Curaçao, the fresh red-brown eggs develop in about three weeks. The eggs then contain a full-grown larva, in the first zoea stage, which is liberated as soon as the egg is brought into the sea. In a number of consecutive nights the females deposit clusters of ripe eggs at the low water line, from which clusters the larvae are carried away by the rising tide. Apart from females with fully developed eggs another, second, wave of animals with freshly laid eggs may be observed, to be followed in some cases by a third wave. In this way groups of larvae in consecutive cycles, are brought to the sea throughout the summer season. The reproductive period lasts from July to about November on the Leeward Group of the Lesser Antilles. A striking feature is the difference in sex ratio between younger and older animals, the relative number of females decreasing with increasing age.

4. Land hermit crabs are frequently exposed to strong evaporation, therefore a good w a t e r m a n a g e m e n t is of primary importance. There are many factors contributing to this maintenance, such as adaptations in anatomy, way of life and behaviour, the powers of detection and uptake of water, the mainly nocturnal life, the possession of a shell which can be closed, the shell water, the urge to seek a suitable micro-habitat, etc. Under constant environmental conditions dehydration always proceeds along the same lines in the same animal. At a temperature of 28°C and a relative air humidity of about 75%, which for Curaçao are the normal conditions, average survival was 8 days, together with water losses of maximally 30% of the initial amount of body water. Dehydration always entails an increase in osmotic value of the body fluids; uptake of water makes the concentration decrease again.

5. In favourable conditions *Coenobita clypeatus* actively stores a great amount of water as s h e l l w a t e r. By applying salt accumulation, washing, mixing or dilution, in relation to the salinity of the available drinking water, the animal effects a rough regulation, in which the shell water is kept more or less constantly at a salinity fluctuating around 32–33‰ S, which is slightly below the value of normal sea water (= 36‰ S). As a rule small animals maintain a slightly higher concentration of shell water than large animals. Air humidity also influences the concentration of shell water in such a way that in a period of drought shell water of a slightly lower concentration is stored. Generally speaking the time factor has a stabilizing influence on the concentration. When in an experiment drinking water of various concentrations is offered, there is a tendency to take up water of ever decreasing salinities, while the total amounts are decreasing too.

6. O s m o r e g u l a t i o n fails in land hermit crabs that do not possess an external environment of water. Due to evaporation and consequently dehydration of the animal the osmoconcentration may reach very extreme, lethal heights, equivalent to over 60‰ S. Dehydrating hermit crabs may be compared to osmoconformers. If, however, such an external environment is actually present, *Coenobita* proves to be a fairly good regulator, trying to keep its body fluid as constant as possible, both in hypotonic and hypertonic media. Generally the role of the external environment is played by the shell water, which serves as an intermediate environment between the available water and the internal medium. As this shell water is already regulated, the osmoregulation proper must be considered as superimposed on that of the shell water. An optimal value for osmoconcentration of the body fluid approaches an equivalent of 34–35‰ S. Non-electrolytes contribute greatly – up to 30% – to the total concentration. Osmoregulation however is mainly brought about by regulation of the electrolytes.

RESUMEN

1. Esta tesis trata sobre varios aspectos de la v i d a, de la ecología, el manejo de agua y la regulación osmótica del cangrejo ermitaño terrestre *Coenobita clypeatus* (Herbst) en Curaçao, Antillas Neerlandesas.

2. Los cangrejos ermitaños terrestres que pertenecen a la familia de Coenobitidae pueden ser considerados como una de las especies más típicamente terrestres de los decápodos. Son característicos de las costas y las islas tropicales. Aunque puedan ser encontrados en una gran variedad de ambientes, los *C. clypeatus*, prefieren regiones con un clima relativamente árido. El a m b i e n t e que forman las Islas de Sotavento es muy apropiado para estos animales. Además de los ejemplares de estos animales que viven en la costa, los llamados "animales costales," hay también otros grupos que viven en el interior, grupos que llamaremos "animales del interior." Estos últimos generalmente son ejemplares más viejos que viven en bien conservados caracoles *Livona pica*; viven en lugares donde hay abundancia de alimentos y de agua dulce o salobre. Aunque existe todavía un fuerte lazo que une estos últimos animales con el mar, se puede decir que han alcanzado un alto nivel de vida terrestre. Gran parte de nuestra investigación fue llevada a cabo con los animales que viven en el interior.

3. En el mes de julio empieza la m i g r a c i ó n relacionada con la reproducción. Para esta fecha tanto los animales adultos como los que todavía no pueden ser considerados como tales, migran a un número determinado de lugares de la costa meridional, probablemente por sendas ya conocidas. Estos parajes se caracterizan por abundancia de cobijo, cantidad suficiente de agua potable y una costa tranquila y accesible. Los animales que tienen más o menos el mismo tamaño suelen agruparse. Se supone que esta época coincide con la fertilización. Poco tiempo después, generalmente alrededor de la primera luna nueva siguiente, se puede notar la presencia de huevos en las hembras. En las circunstancias reinantes en Curaçao, los huevos de un color entre rojo y castaño, maduran dentro de unas tres semanas. Para entonces, cada huevo contiene una larva madura en la primera fase zoea, que sale del huevo tan pronto como éste entra en contacto con el agua del mar. Durante algunas noches consecutivas las hembras, aprovechándose del reflujo, ponen racimos de huevos completamente desarrollados en el fondo de las rocas. El altamar luego se encarga de llevar las larvas. Además de las hembras con huevos desarrollados, se puede observar un segundo grupo, con huevos nuevos, y hasta un tercer grupo. De este modo sucesivos grupos de larvas en diversos ciclos de evolución son depositados en el mar durante el verano. El período reproductivo abarca los meses de julio hasta noviembre, más o menos, en las Islas de Sotavento. Un rasgo muy sorprendente es la diferencia numerativa en sexo entre los animales jóvenes y los viejos. El número de hembras disminuye con la edad.

4. Los cangrejos ermitaños terrestres frecuentemente están expuestos a una importante deshidratación. Por eso un buen m a n e j o d e a g u a es de primordial importancia. Hay muchos factores que contribuyen a ese manejo de agua, tales como adaptaciones anatómicas, modo de vivir, comportamiento, la capacidad de encontrar y de poder abastecerse de agua, una vida casi exclusivamente nocturna, un caracol que se puede cerrar, el agua que puede conservar en el caracol, la preferencia por un micro habitat, etc. En circunstancias ambientales constantes la deshidratación siempre se efectúa según los mismos patrones, tratándose del mismo animal. A una temperatura de 28° centígrados y una humedad relativa del aire de unos 75%, cosas muy comunes en Curaçao, el promedio de supervivencia era de 8 días con una pérdida de agua de a lo sumo 30% de la cantidad inicial. La deshidratación siempre acarrea un incremento del valor osmótico de los líquidos corporales. El tomar agua hace que disminuya nuevamente la concentración.

5. En circunstancias favorables, *Coenobita clypeatus* almacena una buena cantidad de a g u a e n s u c a r a c o l. Acumulando sal, lavándose, mezclando o diluyendo de acuerdo con la salinidad del agua potable disponible, el animal efectúa cierta regulación por medio de la cual el agua en el caracol se mantiene a una salinidad constante de más o menos 32 a 33‰ S, lo cual es algo menos que la salinidad normal del agua del mar (36‰ S). Normalmente los animales pequeños mantienen una concentración algo más alta en el agua que se encuentra dentro de su caracol, que los mayores. La humedad del aire también influye sobre la concentración acuosa en el caracol en tal modo que en períodos de sequía se almacena una cantidad de agua de concentración más baja. En general se puede decir que el factor tiempo ejerce una influencia estabilizadora en la concentración. Cuando durante un experimento, les ofrecimos a los animales agua potable de varias concentraciones decrecientes, disminuyé al mismo tiempo la cantidad total de sales.

6. La r e g u l a c i ó n o s m ó t i c a falla en la vida de los cangrejos ermitaños terrestres que carecen de medio ambiente adecuado. A causa de la evaporación que trae consigo la deshidratación del animal, la concentración osmótica puede alcanzar un grado letal sumamente alto que sobrepasa el 60‰ S. Se puede comparar los cangrejos ermitaños, que se están deshidratando, con animales no-reguladores. Si, por el contrario, se halla presente un medio ambiente favorable, los *Coenobitas* resultan buenos reguladores que tratan de mantener el líquido en su cuerpo a un nivel más constante posible, tanto en un ambiente hipotónico como en el hipertónico.

Generalmente es el agua que se conserva en el caracol la que desempeña el papel de ambiente externo y sirve como enlace entre el aqua disponible y el medio ambiente interno. Cuando el aqua del caracol ha sido regulada, la función osmótica propiamente dicha debe ser considerada como más importante que el contenido acuoso del caracol. El grado óptimo de regulación osmótica en el líquido corporal es de una equivalencia aproximada entre 34 y 35‰ S. Los no-electrólitos contribuyen con un 30% de la concentración total. La regulación osmótica, sin embargo, es ocasionada particularmente por la regulación de los electrólitos.

REFERENCES

ABBOTT, R. T., 1972. The Bermuda "Wilke." *Newsletter Bermuda Biol. Sta. 2, 1*: 2.

BLISS, D. E., 1956. Neurosecretion and the control of growth in a decapod crustacean. In: WINGSTRAND, *Bertil Hanström, Zool. papers in honour of his sixty-fifth birthday*: 56–75. Lund.

BLISS, D. E., 1963. The pericardial sacs of terrestrial Brachyura. In: WHITTINGTON & ROLFE, *Phylogeny and evolution of Crustacea*: 59–78. MCZ Harvard.

BLISS, D. E., 1968. Transition from water to land in decapod crustaceans. *Am. Zoologist 8*: 355–392.

BLISS, D. E. & WANG, S. M. & MARTINEZ, E. A., 1966. Water balance in the land crab, Gecarcinus lateralis, during the intermolt cycle. *Am. Zoologist 6*: 197–212.

BOUSFIELD, E. L., 1968. (Discussion following BLISS, Transition to land in Decapods). *Am. Zoolologist 8*: 393–395.

CARPENTER, A. & LOGAN, W. H., 1945. Hermits don't always live alone. *Natural History A.M.N.H. 54*: 286–287.

CHASE Jr., F. A. & HOBBS Jr., H. H., 1969. *The freshwater and terrestrial decapod crustaceans of the West Indies with special reference to Dominica*. Bull. U.S. N.M. *292*, 258 pp.

EDNEY, E. B., 1957. *The water relations of terrestrial arthropods*, 109 pp. Monogr. Exp. Biol., Cambr. Un. Press.

EDNEY, E. B., 1960. Terrestrial adaptations. In: WATERMAN, *The physiology of Crustacea*, I: 367–388. Acad. Press, N.Y. & London.

EMERY, A. R., 1972. Eddy formation from an oceanic island: ecological effects. *Car. J. Sci. 12*; 121–128.

FUGLISTER, F. C., 1960. *Atlantic Ocean atlas of temperature and salinity* ..., 209 pp. Woods Hole.

GRIMM, A. S., 1969. *Osmotic and ionic regulation in the shrimps Crangon vulgaris Fabricius and Crangon allmanni Kinahan*, 88 pp. Thesis, Glasgow.

GROSS, W. J., 1955. Aspects of osmotic regulation in crabs showing the terrestrial habit. *Am. Naturalist 89*: 205–222.

GROSS, W. J., 1957. An analysis of response to osmotic stress in selected decapod Crustacea. *Biol. Bull. 112*: 43–62.

GROSS, W. J., 1964. Water balance in anomuran land crabs on a dry atoll. *Biol. Bull. 126*: 54–68.

GRUBER, M. & SHOUP, J. B., 1969. Crabs move ashore. *Sea Frontiers 15*: 364–375.

HARMS, J. W., 1929. Die Realisation von Genen und die Consecutive Adaption, I. Phasen in der Differenzierung der Anlagenkomplexe und die Frage der Landtierwerdung. *Z. Wiss. Zool. 133*: 211–397.

HARMS, J. W., 1932. Die Realisation von Genen und die Consecutive Adaption, II. Birgus latro L. als Landkrebs und seine Beziehungen zu den Coenobiten. *Z. Wiss. Zool. 140*: 167–290.

HARMS, J. W., 1937. Lebenslauf und Stammesgeschichte des Birgus latro L. von den Weihnachtsinseln. *Z. Naturwiss. Jena 71*: 1–34.

HAZLETT, B. A., 1966. Observations on the social behavior of the land hermit crab, Coenobita clypeatus (Herbst). *Ecology 47*: 316–317.

HAZLETT, B. A., 1966. Social behavior of the Paguridae and Diogenidae of Curaçao. *Stud. fauna Curaçao 23*: 1–143.

HOHENDORF, K., 1963. Der Einflusz der Temperatur auf die Salzgehaltstoleranz und Osmoregulation von Nereis diversicolor O.F. Muell. *Kieler Meeresf. 19*: 196–218.

KALBER, F. A. & COSTLOW, J. D., 1968. Osmoregulation in the larvae of the land-crab, Cardisoma guanhumi Latreille. *Am. Zoologist 8*: 411–416.

PALMER, J. D., 1971. Comparative studies of circadian locomotory rhythms in four species of terrestrial crabs. *Am. Midland Nat. 85*: 97–107.

PEARSE, A. S., 1916. An account of the Crustacea collected by the Walker Expedition to Santa Marta, Colombia. *Proc. U.S. Nat. Mus. 49*: 531–556.

PEARSE, A. S., 1929. Observations on certain littoral and terrestrial animals at Tortugas, Florida, with special reference to migrations from marine to terrestrial habitats. *Papers Tortugas Lab.* 24: 205–223.

PROSSER, C. L. & BROWN, F. A., 1962. *Comparative animal physiology* (2nd. ed.), 688 pp. Saunders, Phila. & London.

PROVENZANO Jr., A. J., 1959. The shallow-water hermit crabs of Florida. *Bull. Mar. Sci. Gulf & Carib. 9*: 349–420.

PROVENZANO Jr., A. J., 1962. The larval development of the tropical land hermit Coenobita clypeatus (Herbst) in the laboratory. *Crustaceana 4*: 207–228.

PHILIPS, 1963. *Universele 12-kanalen recorder PR 3210 U/OO*. [Service manual]. Eindhoven.

SEURAT, L. G., 1904. Observations biologiques sur les cénobites (Coenobita perlata Edwards). *Bull. Mus. Hist. Nat. 10*: 238–242.

SPAARGAREN, D. H., 1971. Aspects of the osmotic regulation in the shrimps Crangon crangon and Crangon allmanni. *Neth. J. Sea Res. 5*: 275–335. Thesis, Amsterdam.

Statistiek van de meteorologische waarnemingen van de Nederlandse Antillen. 1963–1968. Curaçao.

STOFFERS, A. L., 1956. *The vegetation of the Netherlands Antilles.* Publ. Found. Sci. Res. Sur. & N.A. *15*: 142 pp. Thesis, Utrecht.

Terrestrial adaptations in Crustacea. [Symposium, 27–29.XII.1967]. *Am. Zoologist 8*, 1968: 307–685.

TUCKER ABBOTT, R., 1972. The Bermuda "Wilke." *Newsletter Bermuda Biol. Sta. 2, 1*: 2.

VERRILL, A. E., 1908. Decapod Crustacea of Bermuda. 1. Brachyura and Anomura. *Trans. Conn. Acad. Sci. 13*: 299–474.

VERWEY, J., 1957. A plea for the study of temperature influence on osmotic regulation. *Ann. Biol. 33*: 129–149.

VÖLKER, L., 1965. *Experimentelle Untersuchungen zur Ökologie des Landeinsiedlerkrebses Coenobita scaevola Forskål am Roten Meer,* 115 pp. typescript. Thesis, Techn. Hochschule Darmstadt.

VÖLKER, L., 1967. Zur Gehäusewahl des Land-einsiedlerkrebses Coenobita scaevola Forskål vom Roten Meer. *J. exp. mar. Biol. Ecol. 1*: 168–190.

WIENS, H. J., 1962. *Atoll environment and ecology,* 532 pp. Yale Univ. Press.

WILDE, P. A. W. J. DE, 1969. Research on terrestrial crabs in Curaçao, Netherlands Antilles. *WOTRO report for the year 1968*: 38–42. The Hague.

WILDE, P. A. W. J. DE, 1972. Soldaatjeskrabben. *Stinapa 6*: 37–47. Curaçao.

WOLVEKAMP, H. P. & WATERMAN, T. H., 1960. Respiration. In: WATERMAN, *The physiology of Crustacea* I: 35–100. Acad. Press N.Y. & London.

WÜST, G., 1964. *Stratification and circulation in the Antillen-Caribbean basins* (Part 1), 199 pp. Columbia Univ. Press.

SAMENVATTING

1. Deze publicatie gaat over verschillende aspecten van de l e v e n s g e-s c h i e d e n i s, oecologie, waterhuishouding en osmoregulatie van de West-indische Landheremietkrab *Coenobita clypeatus* (Herbst) op Curaçao, Neder-landse Antillen.

2. Landheremietkrabben, behorend tot de familie der Coenobiti dae, moe-ten tot de meest terrestrische vormen onder de Decapoden gerekend worden. Het zijn kenmerkende dieren van tropische kusten en eilanden. Ofschoon *C. clypeatus* in een verscheidenheid van l a n d s c h a p p e n aangetroffen kan worden, hebben woongebieden met een relatief droog klimaat de voorkeur. Habitats als op de Benedenwindse Eilanden kunnen in dit opzicht represen-tatief genoemd worden. Naast populaties van langs de kust levende dieren, "coastal animals," worden meer landinwaarts de z.g. "inland animals" on-derscheiden. Over het algemeen bestaat deze laatste groep uit oudere exem-plaren in goed passende en gave schelpen van de soort *Livona pica*, die zich in de voedselrijke habitats, met zoet of brak water, hebben kunnen vestigen en handhaven. Niettegenstaande ook deze dieren een nauwe binding met de zee onderhouden, worden zij beschouwd een geavanceerder stadium van land-leven bereikt te hebben. Het meeste onderzoek werd aan "inland animals" verricht.

3. In juli begint de v o o r t p l a n t i n g s-m i g r a t i e, waarbij volwassen en onvolgroeide dieren – vermoedelijk langs traditionele wegen – naar een be-perkt aantal paaiplaatsen aan de zuidkust trekken. Deze plaatsen worden gekenmerkt door de aanwezigheid van goede schuilplaatsen, geschikt drink-water en een rustig en accessibel kustgedeelte. Veelal verzamelen dieren van ongeveer een zelfde grootte-klasse zich op afzonderlijke plaatsen. Aangeno-men wordt dat hier de bevruchting plaats vindt. Spoedig daarna, meestal omstreeks volle maan, kunnen de eerste vrouwtjes met jonge, roodbruine eieren, vastgehecht aan de pleopoden van het abdomen, waargenomen wor-den. Onder de op Curaçao voorkomende omstandigheden ontwikkelen de eieren zich in ongeveer drie weken tot een volgroeide larve in het eerste zoea stadium. Klompjes rijpe eieren worden in een aantal opeenvolgende nachten door de vrouwtjes op de laagwaterlijn verspreid. In aanraking met water barsten de eieren terstond open; de larven worden met het vloedwater mee-gevoerd. Naast vrouwtjes met geheel ontwikkelde eieren wordt nu opnieuw een tweede groep dieren met pas gelegde eieren waargenomen, die op zijn beurt door een derde groep gevolgd kan worden. In de loop van het voort-plantingsseizoen worden aldus in een aantal opeenvolgende golven groepen larven aan de zee toevertrouwd. De voortplantingsperiode loopt op de Be-nedenwindse Eilanden van juli tot omstreeks november. Opvallend zijn de verschillende verhoudingen in sexe tussen jonge en oude dieren, zodanig dat het aantal vrouwtjes bij oudere dieren relatief sterk afneemt.

4. Landheremieten staan voortdurend aan sterke evaporatie bloot, zodat een goede w a t e r h u i s h o u d i n g van primaire betekenis is. Tal van adap-taties in bouw, levenswijze en gedrag, de fijne detectie- en opnamemogelijk-heden van water, de overwegend nachtelijke levenswijze, het bezit van een goed afsluitbare schelp, het schelpwater, en het opzoeken van de juiste micro-

habitat leveren hiertoe een belangrijke bijdrage. Onder gelijke omstandigheden verloopt het waterverlies bij een bepaald dier steeds op dezelfde wijze. Bij een temperatuur van 28°C en een relatieve luchtvochtigheid van ongeveer 75%, dus onder voor Curaçao normale omstandigheden, werden gemiddelde overlevingstijden van 8 dagen gemeten, hetgeen gepaard ging met waterverliezen van maximaal 30% van de initiële hoeveelheden lichaamswater. Dehydratie brengt steeds een stijging van de osmotische waarde van de lichaamsvloeistoffen met zich mee; opname van water doet de concentratie weer dalen.

5. *Coenobita clypeatus* slaat onder gunstige omstandigheden actief grote hoeveelheden water als s c h e l p w a t e r op. Door in afhankelijkheid van het zoutgehalte van het beschikbare drinkwater op adequate wijze zoutaccumulatie, doorspoeling, menging of verdunning toe te passen, geraakt het dier tot een grove regulatie, waarbij de saliniteit van het schelpwater min of meer constant gehouden wordt op een waarde schommelend rond 32–33‰ S, dus iets beneden de waarde van het normale zeewater (= 36‰ S). Kleine dieren onderhouden doorgaans een wat hogere concentratie van het schelpwater dan grote dieren. De vochtigheid van de lucht blijkt eveneens van invloed op het schelpwater te zijn, zodanig dat in een drogere omgeving water van een lagere saliniteit vastgehouden wordt. In het algemeen heeft de factor tijd een stabiliserende invloed op de concentratie. Wanneer onder experimentele omstandigheden drinkwater van verschillende concentratie aangeboden wordt, bestaat de tendens om water van steeds lager zoutgehalte op te nemen, waarbij de totaal opgenomen hoeveelheden afnemen.

6. Bij landheremietkrabben, die niet beschikken over een uitwendig waterig milieu, faalt de o s m o r e g u l a t i e, waardoor tengevolge van evaporatie en de daarmee gepaard gaande dehydratie van het dier de osmoconcentratie tot zeer extreme, letale waarden, equivalent met ruim 60‰s, kan stijgen. Uitdrogende heremieten kunnen met osmoconformers vergeleken worden. Wanneer echter een dergelijk extern milieu wel aanwezig is, manifesteren Coenobitas zich als tamelijk goede regulators, die er naar streven de concentratie van de lichaamsvloeistof zo constant mogelijk te houden, zowel in hypo- als in hypertonische media.

Veelal vervult het schelpwater de rol van het externe milieu, waardoor het als intermediair milieu tussen het beschikbare drinkwater en het interne milieu geschakeld wordt. Gezien de regulatie van het schelpwater, moet de eigenlijke osmoregulatie van het dier als gesuperponeerd op die van het schelpwater gedacht worden. Een optimum waarde voor de osmoconcentratie van de lichaamsvloeistof ligt bij een equivalent van 34–35‰ S. Niet-electrolieten leveren een aanzienlijke bijdrage – tot 30% – aan de totale concentratie. Osmoregulatie geschiedt evenwel hoofdzakelijk door regulatie van de electrolieten.

PLATE I

Ia. 'Oostpunt' estate on Curaçao, as seen from the East towards the limestone escarpments of Duivelsklip (62 m, left) and Tafelberg (196 m, centre). A heavily eroded area with an open vegetation consisting mainly of thorny shrubs, with cactuses, woody herbs, and small trees. (Phot. P. Wagenaar Hummelinck, 28.X. 1963).

Ib. A large specimen of *Coenobita clypeatus* (Herbst) peeping from its *Livona*-shell. (Phot. M. Arnoldo Broeders).

PLATE II

IIa. Well 1: A characteristic coastal habitat of *Coenobita*, showing a concrete water through fed by a wind mill, at the base of a limestone escarpment on the north coast of East-Curaçao. The area is barren, and only scantily covered with *Opuntia wentiana* and low shrubs.

IIb. Punta Tera: A characteristic inland habitat of *Coenobita*, showing two concrete water basins fed by a wind mill. The area is covered with shady trees, incl. *Acacia tortuosa*, *Hippomane mancinella* and *Swietenia mahogany*.

PLATE III

IIIa. Aerial view of the Lagoon area along the south coast of East-Curaçao, showing part of Lagoon A (top-left), Lagoon B, and Lagoon C. Lagoon B is separated by the Landbridge (arrow) from Lagoon C. The semi-circular mudbank in Lagoon C is the so-called Delta.

IIIb. The Landbridge separating Lagoons B and C as seen towards the East. Note the desolate surroundings, with coral stone boulders, and a few bushes of *Conocarpus erecta* which represent the main hiding places for *Coenobita*.

PLATE IV

IVa. *Coenobita clypeatus* in shell of *Livona pica*, crawling around with extended antennae.

IVb. *Coenobita clypeatus*, removed from its shell.

PLATE V

Va. Male *Coenobita clypeatus* showing spermatophores protruding from the paired gonopores.

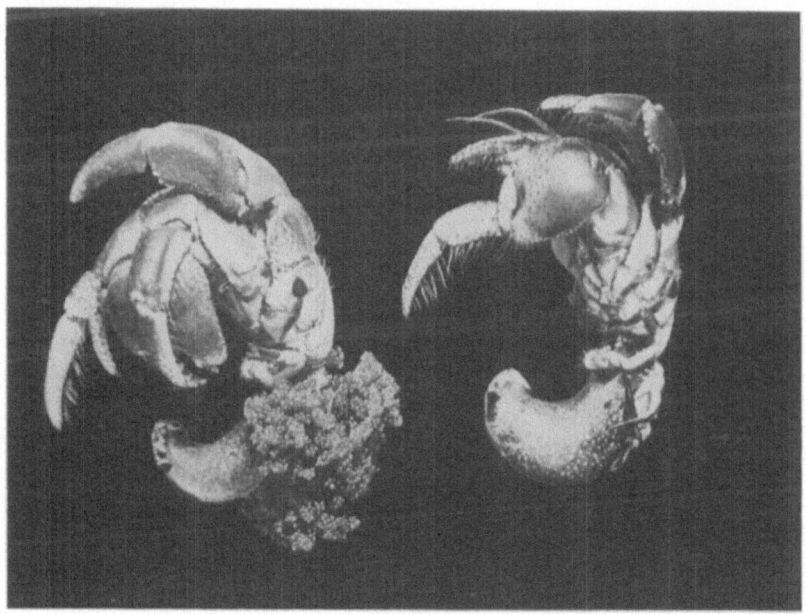

Vb. Female *Coenobita clypeatus*, one of them with egg clusters attached to the hairs of the pleiopods.

PLATE VI

VIa. A most unsuitable type of 'shell', accepted by a *Coenobita* in a bungalow at Savaneta, Aruba. (Courtesy Lago Oil & Transport Co. Ltd., 1961).

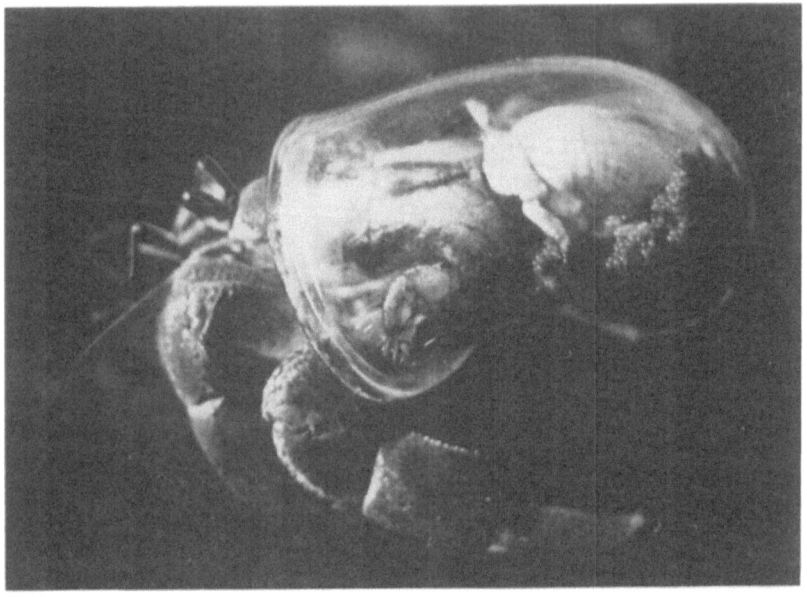

VIb. A glass-shell, accepted by a female *Coenobita* during experiments.

STELLINGEN

1. Heremietkrabben uit de Familie der Coenobitidae kunnen beschouwd worden als de voorlopers van een toekomstige succesvolle groep landdieren
   Dit proefschrift.

2. Transpiratie als een mechanisme voor temperatuursregulatie bij landkrabben moet als weinig zinvol beschouwd worden.
   D. Bliss (1968), Am. Zool. 8: 364–368.

3. De voortplantingsmigratie van *Coenobita clypeatus* gaat langs traditionele routes, waarbij jonge en onervaren dieren door oude dieren geleid worden.
   Dit proefschrift.

4. De combinatie van benthische micro-algen, detritus en bacteriën vormt het voedsel van deposit feeders in de estuariene gebieden.
   R. C. Newell (1973), Biology of intertidal animals. Paul Elek Sc. Books, Londen: 238–263.

5. Bij de meting van het energieverbruik door mariene organismen in het kader van productieonderzoek wordt onvoldoende rekening gehouden met het gedrag en de levenswijze van de dieren.
   P. A. W. J. de Wilde (1973), Neth. J. Sea Res. 6: 157–162.

6. De resultaten van de experimenten van Hesz vinden met het exosomenmodel van Fox en Yoon (1970) een redelijke verklaring.
   D. Hesz (1972), Naturwissenschaften 59: 348–355.

7. Het CARMABI moet in de eerste plaats als een instituut voor oecologisch veldonderzoek gezien worden.

8. Er dient rekening mee gehouden te worden dat de veranderingen in soortensamenstelling en talrijkheid van evertebraten in Noord- en Waddenzee toegeschreven kunnen worden aan andere dan natuurlijke oorzaken.

9. In het Nederlandse deel van de Waddenzee is plaats voor een zeehondenbestand van ongeveer 2000 dieren.

10. De wettelijk toegestane raaptijd van kievitseieren moet ingekort worden tot 7 april; die van andere weidevogels dient geheel gesloten te worden.

11. In verband met ontoelaatbare geluidsoverlast en hinder moeten zo spoedig mogelijk afdoende maatregelen getroffen worden om de snelle expansie van de z.g. kleine luchtvaart boven het Waddengebied in te perken.

12. Nimrod is dood.

P. A. W. J. de Wilde, november 1973